The Man from the Future

The Visionary Ideas of John von Neumann

ANANYO
BHATTACHARYA

W. W. NORTON & COMPANY
Celebrating a Century of Independent Publishing

First published as a Norton paperback 2023

Printed in the United States of America
First published as a Norton paperback 2023

Manufacturing by Lakeside Book Company
Production manager: Devon Zahn

The Library of Congress has catalogued the hardcover edition of this book as follows:

Library of Congress Cataloging-in-Publication Data

Names: Bhattacharya, Ananyo, author.
Title: The man from the future : the visionary life of John von Neumann / Ananyo Bhattacharya.
Description: First American edition. | New York : W. W. Norton & Company, 2022. | Includes bibliographical references and index.
Identifiers: LCCN 2021050487 | ISBN 9781324003991 (hardcover) | ISBN 9781324004004 (epub)
Subjects: LCSH: Von Neumann, John, 1903–1957. | Mathematicians—Hungary—Biography. | Mathematicians—United States—Biography.
Classification: LCC QA29.V66 B43 2022 | DDC 510.92 [B]—dc23/eng/20211201
LC record available at https://lccn.loc.gov/2021050487

ISBN 978-1-324-05050-6 pbk.

W. W. Norton & Company, Inc., 500 Fifth Avenue, New York, N.Y. 10110
www.wwnorton.com

W. W. Norton & Company Ltd., 15 Carlisle Street, London W1D 3BS

1 2 3 4 5 6 7 8 9 0

Praise for *The Man from the Future*

"Vivid. . . . [*The Man from the Future* is] devoted to exploring the ideas and technological inquiries [von Neumann] inspired."
—Jennifer Szalai, *New York Times Book Review*

"[Ananyo] Bhattacharya both begins and concludes this impressive biography of John von Neumann by celebrating his contribution to the 'march of ideas.'" —Francis P. Sempa, *New York Journal of Books*

"Bhattacharya tells the story tremendously well, situating von Neumann's work—in fields from quantum mechanics to game theory to cellular automata—as comfortably as I've ever seen it done. He's also good at deadpan humor." —David Bodanis, *Financial Times* (UK)

"[Bhattacharya] is a first-class science writer with an impeccable pedigree embracing stints at the *Economist* and *Nature*, and he does the best job I have seen of explaining the significance of von Neumann's work across many different fields. . . . [A] fine tribute to [von Neumann's] genius and his contributions to science."
—John Gribbin, *Literary Review* (UK)

"[An] agile, intelligent, intellectually enraptured account of von Neumann's life." —Simon Ings, *Sunday Telegraph* (UK)

"Impressively clear brief accounts of quantum mechanics, Gödel's theorem, and the physics of explosives. . . . Bhattacharya manages to cover this dazzling range of ideas clearly and compellingly."
—Jon Turney, *Arts Desk* (UK)

"Bhattacharya provides the historical context for this moment on the technological timeline with a professional historian's authority. Brilliant." —Nick Smith, *Engineering & Technology*

"Sharp, expansive. . . . A salient portrait of one of the most electrifying and productive scientists of the past century." —*Kirkus Reviews*

"Any future intelligence capable of sending a representative back in time to help invent itself will be intelligent enough to conceal this from us. Ananyo Bhattacharya's *The Man from the Future* is therefore unable to confirm this suggestion, but much else about John von Neumann's presence in the twentieth century is revealed along the way."
—George Dyson, author of *Turing's Cathedral*

"Despite his central contributions to the theory of computation, economics, logic, complexity, and quantum physics, somehow John von Neumann never became a household name to rival Einstein and Feynman. Ananyo Bhattacharya's biography deserves to change that. Consistently clear and careful without sacrificing elegance or accessibility, it does full justice to this legendary figure of twentieth-century science."
—Philip Ball, author of *Beyond Weird*

"An engaging and fascinating book that blends science and history. I loved it."
—Paul Davies, author of *The Demon in the Machine*

"John von Neumann was one of the deepest and broadest thinkers of the twentieth century—but also a delight to encounter. According to one admirer, listening to him was 'mental champagne.' Now we have a biography about which you can say the same. This is a sparkling book, with an intoxicating mix of pen-portraits and grand historical narrative. . . . [A] staggering achievement."
—Tim Harford, author of *How to Make the World Add Up*

"More than just a biography, *The Man from the Future* elucidates the breathtaking scientific progress in the mid-twentieth century, skillfully woven together in the story of one man, John von Neumann."
—Sabine Hossenfelder, author of *Lost in Math*

"A gripping tale of the most significant mathematical, scientific, and geopolitical events of the early twentieth century. Bhattacharya's storytelling seamlessly weaves together the science, the vibrant social and historical context, and the private idiosyncrasies of John von Neumann and the fascinating geniuses around him, without mythologizing."
—Andrew Steele, author of *Ageless*

To geeks and nerds everywhere,
but especially for the three closest to me.

'If people do not believe that mathematics is simple, it is only because they do not realize how complicated life is.'

John von Neumann

Contents

Introduction: Who Was John von Neumann?

> *'Von Neumann would carry on a conversation with my three-year-old son, and the two of them would talk as equals, and I sometimes wondered if he used the same principle when he talked to the rest of us.'*
>
> *Edward Teller, 1966*

Call me Johnny, he urged the Americans invited to the wild parties he threw at his grand house in Princeton. Though he never shed a Hungarian accent that made him sound like horror-film legend Bela Lugosi, von Neumann felt that János – his real name – sounded altogether too foreign in his new home. Beneath the bonhomie and the sharp suit was a mind of unimaginable brilliance.

At the Institute for Advanced Study in Princeton, where he was based from 1933 to his death in 1957, von Neumann enjoyed annoying distinguished neighbours such as Albert Einstein and Kurt Gödel by playing German marching tunes at top volume on his office gramophone. Einstein revolutionized our understanding of time, space and gravity. Gödel, while no celebrity, was equally revolutionary in the field of formal logic. But those who knew all three concluded that von Neumann had by far the sharpest intellect. His colleagues even joked that von Neumann was descended from a superior species but had made a detailed study of human beings so he could imitate them perfectly.

As a child, von Neumann absorbed Ancient Greek and Latin, and spoke French, German and English as well as his native Hungarian. He devoured a forty-five-volume history of the world and was able to recite whole chapters verbatim decades later. A professor of Byzantine history who was invited to one of von Neumann's parties said he would come only if it was agreed they would not discuss the subject. 'Everybody thinks I am the world's greatest expert in it,' he told von Neumann's wife, 'and I want them to keep on thinking that.'

The principal focus of von Neumann's incredible brain, however, was neither linguistics nor history but mathematics. Mathematicians often describe what they do as a sort of noble game, the object of which is to prove theorems, divorced from any real application. That is often true. But maths is also the language of the sciences – the most powerful tool we have for understanding the universe. 'How can it be that mathematics,' asked Einstein, 'being after all a product of human thought which is independent of experience, is so admirably appropriate to the objects of reality?'[1] No one has come up with a definitive answer to that question. Since antiquity, however, mathematicians with a talent for its application have, like von Neumann, understood that they have a path to wealth, influence and the power to transform the world. Archimedes spent time on otherworldly pursuits such as finding a new way to approximate the number pi. But the war machines he designed to exacting mathematical principles, such as a giant claw that could pluck ships from the sea, frustrated for a time the Roman army.

The mathematical contributions von Neumann made in the mid-twentieth century now appear more eerily prescient with every passing year. To fully understand the intellectual currents running through our century – from politics to economics, technology to psychology – one has to understand von Neumann's life and work in the last. His thinking is so pertinent to the challenges we face today that it is tempting to wonder if he was a time traveller, quietly seeding ideas that he knew would be needed to shape the Earth's future.

Born in 1903, von Neumann was just twenty-two years old when he helped to lay the mathematical foundations of quantum mechanics. He moved to America in 1930 and, realizing early on that war was looming, studied the mathematics of ballistics and explosions. He lent his expertise to the American armed forces and the Manhattan Project: among the scientists at Los Alamos who developed the atomic bomb, it was von Neumann who determined the arrangement of explosives that would be required to detonate the more powerful 'Fat Man' device by compressing its plutonium core.

The same year he joined the Manhattan Project, von Neumann was finishing, with the economist Oskar Morgenstern, a 640-page treatise on game theory – a field of mathematics devoted to understanding conflict and cooperation. That book would change economics, make game theory integral to fields as disparate as political science, psychology and evolutionary biology and help military strategists to think about when leaders should – and should not – push the nuclear button. With his unearthly intelligence and his unflinching attitude to matters of life and death, von Neumann was one of a handful of scientists who inspired the iconic Stanley Kubrick character Dr Strangelove.

After the atom bombs he helped to design were dropped on Hiroshima and Nagasaki, von Neumann turned his efforts to building possibly the world's first programmable electronic digital computer, the ENIAC. Initially his aim was to calculate whether or not it would be possible to build a more powerful bomb – the hydrogen bomb. He then led the team that produced the first computerized weather forecast. Not content with computers that merely calculated, von Neumann showed during a lecture in 1948 that information-processing machines could, under certain circumstances, reproduce, grow and evolve. His automata theory inspired generations of scientists to try and build self-replicating machines. Later, his musings on the parallels between the workings of brains and computers helped to trigger the birth of artificial intelligence and influenced the development of neuroscience.

Von Neumann was a pure mathematician of extraordinary ability. He established, for example, a new branch of mathematics, now named after him, that was richly productive: half a century later, Vaughan Jones won the Fields Medal – often called the Nobel Prize of maths – for his work exploring one tiny aspect of it. But mere intellectual puzzles, no matter how profound, were not enough for him. Von Neumann constantly sought new practical fields to which he could apply his mathematical genius, and he seemed to choose each one with an unerring sense of its potential to revolutionize human affairs. 'As he moved from pure mathematics to physics to economics to engineering, he became steadily less deep and steadily more important,' observed von Neumann's former colleague, mathematical physicist Freeman Dyson.[2]

When he died, aged just fifty-three, von Neumann was as famous as it is possible for a mathematician to be. The writer William S. Burroughs claimed von Neumann's game theory inspired some of his bizarre literary experiments, and he is name-checked in the novels of Philip K. Dick and Kurt Vonnegut. Since then, however, von Neumann has, compared with his august Princeton associates, faded from view. Caricatured as the coldest of cold warriors, and with wide-ranging contributions that are almost impossible to summarize, when von Neumann is remembered it is largely for his legendary feats of mental gymnastics. Yet his legacy is omnipresent in our lives today. His views and ideas, taken up by scientists, inventors, intellectuals and politicians, now inform how we think about who we are as a species, our social and economic interactions with each other and the machines that could elevate us to unimaginable heights or destroy us completely. Look around you and you will see Johnny's fingerprints everywhere.

I

Made in Budapest

A genius is born and bred

'Von Neumann was addicted to thinking, and in particular to thinking about mathematics.'

Peter Lax, 1990

The scientists and technicians working on America's secret atom bomb project at Los Alamos during the 1940s called them the 'Martians'. The joke was that with their strange accents and exceptional intellects, the Hungarians among them were from some other planet.

The Martians themselves differed on why one small country should churn out so many brilliant mathematicians and scientists. But there was one fact upon which they were all agreed. If they came from Mars, then one of their number came from another galaxy altogether. When the Nobel Prize-winning physicist and Martian Eugene Wigner was asked to give his thoughts on the 'Hungarian phenomenon', he replied there was no such thing. There was only one phenomenon that required any explanation. There was only one Johnny von Neumann.

Neumann János Lajos (in English, John Louis Neumann – the surname comes first in Hungarian) was born in sparkling Belle Epoque Budapest on 28 December 1903. Created when the old capital of Buda was merged with the nearby cities of Óbuda and Pest in 1873, Budapest was thriving. The Hungarian parliament building on the banks of the Danube was the largest in the world and the grand Beaux Arts stock exchange palace was unrivalled in Europe. Beneath Andrássy Avenue, a glorious boulevard lined with neo-renaissance mansions, ran one of the world's first electrified underground railway lines.

Intellectuals flocked to the coffee houses (the city boasted more than 600), and the acoustics of the opera house, also built around this time, are still considered to be among Europe's finest.

Johnny – his family and friends in Hungary called him Jancsi (pronounced *yan-shi*), a diminutive version of János – was the first of three sons born to Miksa (often translated to Max) and Margit (Margaret), educated, well-to-do parents plugged into the Hungarian capital's dazzling intellectual and artistic life. His brother Mihály (Michael) followed in 1907, and Miklós (Nicholas) in 1911. The family lived in an eighteen-room apartment on the top floor of 62 Vaczi Boulevard.[1]

The ground floor of the home was occupied by the sprawling salesrooms of Kann-Heller, a hardware firm founded by Margaret's father, Jacob Kann, and his partner. Kann-Heller had sold farm machinery, then successfully pioneered catalogue sales in Hungary, much as Sears had done earlier in the United States. The Heller family had the run of the whole first floor. The second and third floors were occupied by Kann's four daughters and their families. Today, on the corner of the building, next to the entrance to the offices of an insurance firm, is a plaque honouring 'one of the most outstanding mathematicians of the 20th century'.

In 1910, a quarter of Budapest's population, and more than half of its doctors, lawyers and bankers, were Jewish, as were many of those involved in the city's thriving cultural scene. In that success, some sought to find conspiracy. The supposed domination of the city by Jews led Karl Lueger, the firebrand populist mayor of Vienna, to call Austria-Hungary's twin capital 'Judapest'. Lueger's racist rhetoric would inspire a young homeless Adolf Hitler, kicking around Vienna after being denied admission to the city's Academy of Fine Arts.

The brunt of Jewish immigration to Hungary had taken place in the last two decades of the nineteenth century. Many, in search of jobs, had settled in rapidly growing Budapest. Jews did not face pogroms as they did in Russia, and the anti-Semitism that for generations has run deep and strong through Europe was, if not altogether absent, at least not usually sanctioned by the government. 'Respectable opinion, including most of the aristocracy and gentry, rejected anti-Semitism,' notes Hungarian-American historian John Lukacs.[2]

Still, for all their prosperity and happiness, the Neumanns were, like many Jews in the Austro-Hungarian Empire, haunted by the anxiety that the good times would not last. Though the dozens of ethnic groups living within its borders were ostensibly united under the popular emperor in Vienna, and by an economic logic that allowed the free movement of goods and services across a huge swathe of southeastern Europe, differences sometimes came to the fore. Robert Musil, one of the empire's many great writers, said that the numerous internecine conflicts 'were so violent that they several times a year caused the machinery of State to jam and come to a dead stop. But between whiles, in the breathing-spaces between government and government, everyone got on excellently with everyone else and behaved as though nothing had ever been the matter'.[3]

Despite the febrile atmosphere of Austria-Hungary, it would not be internal divisions but the First World War that would precipitate the empire's downfall. By 1910, Max had sensed the darkening mood in Europe and wanted his sons' education to prepare them for the worst. Children did not then start school in Hungary until the age of ten, but affluent Budapestian families had no problems finding nursemaids, governesses or tutors. Max emphasized foreign languages, reasoning that his sons would then be able to make themselves understood no matter where they were or who happened to be in charge. So six-year-old Jancsi learned French from Mademoiselle Grosjean and Italian from Signora Puglia. Between 1914 and 1918, the brothers were also taught English by Mr Thompson and Mr Blythe. Though held as enemy aliens in Vienna at the start of the war, Max, a man of influence, had 'no difficulties in having their place of "internment" officially moved to Budapest'.[4] Max also insisted the boys learn Ancient Greek and Latin. 'Father', Nicholas recalled in his memoirs, 'believed in the life of the mind.'[5]

Jancsi was a formidable mental calculator even as a child.[6] Some sources suggest that he could multiply two eight-digit numbers together in his head when he was six.[7] These abilities, remarkable enough to astonish his early tutors, may have been partly inherited from his maternal grandfather. Though Jacob Kann had no formal education beyond secondary school, he could add or multiply numbers into the millions. Von Neumann would recall his twinkly-eyed grandfather's mental gymnastics with pride when he was older, but he admitted he was never quite able to match them himself.

The eldest of the von Neumann brothers did not, however, shine in all things. He never, for instance, mastered a musical instrument. Puzzled that young Jancsi only ever played scales on the cello, his family investigated to find that the five-year-old had taken to propping up books on his music stand so he could read while 'practising'. At chess, a game often associated with mathematical ability, he was middling.[8] Despite developing various 'systems' that he thought would inevitably lead to a win, he lost consistently against his father even as a teenager.

Equally, von Neumann had no interest in sport and, barring long walks (always in a business suit), he would avoid any form of vigorous physical exercise for the rest of his life. When his second wife, Klári, tried to persuade him to ski, he offered her a divorce. 'If being married to a woman, no matter who she was, would mean he had to slide around on two pieces of wood on some slick mountainside,' she explained, 'he would definitely prefer to live alone and take his daily exercise, as he put it, "by getting in and out of a pleasantly warm bathtub".'[9]

Intellectually, life at home was as stimulating as any child prodigy could wish. Max, a doctor of law turned investment banker, bought a library from the estate of a wealthy family when the boys were young. He converted a room in the apartment to house the collection, with shelves from floor to ceiling. This was where Jancsi would make his way through the library's centrepiece, the *Allgemeine Geschichte*, a massive history of the world edited by the German historian Wilhelm Oncken, which began in Ancient Egypt and concluded with a biography of Wilhelm I, the first German emperor, commissioned by the Kaiser himself. When von Neumann became embroiled in American politics after he emigrated, he would sometimes avoid arguments that were threatening to become too heated by citing (sometimes word for word) the outcome of some obscurely related affair in antiquity that he had read about in Oncken as a child.

The children's education often continued over lunch and dinner, when they were encouraged to present a particular topic that had caught their attention earlier in the day. Once, for example, Nicholas

Jancsi in a sailor's suit, aged seven. *Courtesy of Marina von Neumann Whitman.*

read up on the poetry of Heinrich Heine, sparking a discussion on how anti-Semitism would affect them in future. Heine was born into a Jewish family but reluctantly converted to Christianity, 'the ticket of admission into European culture', in an effort to boost his career. Frank debates such as this may have helped Jancsi recognize early on the dangers posed by National Socialism.

Jancsi's mealtime seminars were often on scientific subjects. He noted that babies of different nationalities learned their native languages in about the same length of time. What then, he asked, is the brain's primary language? How does the brain communicate with itself? This is a question that he would continue to wrestle with, even on his deathbed. On another occasion, he wondered whether the spiral cavity of the inner ear known as the cochlea was sensitive only to the component frequencies of sound (and their respective volumes) or the shape of the sound wave as a whole.[10]

Max, who would lunch at home before returning to the office in the afternoon, would share his investment decisions and ask his sons' opinions. Occasionally, Max would bring home more tangible evidence of the companies in which he was investing. When he financed a newspaper business, he brought back pieces of metal type, and the ensuing discussion centred on the printing press. Another venture to win Max's support was the Hungaria Jacquard Textile Weaving Factory, an importer of automated looms.[11] Invented in the early nineteenth century by the Frenchman Joseph Marie Charles (known as 'Jacquard'), these devices could be 'programmed' with punched cards. 'It probably does not take much imagination to trace this experience to John's later interest in punched cards!' notes Nicholas.[12]

Guests at the Neumanns' table were also contributing to the prodigy's academic development. Businessmen from all over Europe would find themselves politely pummelled by questions from Max's sons, who were allowed to sit in on working dinners. Other regular visitors included the psychoanalyst Sándor Ferenczi, a close associate of Sigmund Freud, whose conversation may have helped to form Johnny's later thoughts on the parallels between computers and brains. Physicist Rudolf Ortvay would call, fresh from studies at the University of Göttingen, the world's leading centre of mathematics and soon to be pivotal in the development of the new quantum mechanics. Ortvay would keep up a correspondence with Jancsi throughout his life. Another frequent guest, Lipót (Leopold) Fejér, occupied a chair of mathematics at the University of Budapest. He was soon to be one of several inspiring professors charged with giving the boy extracurricular maths lessons.

After 1910, Max became an economic adviser to the Hungarian government, a role that would rapidly propel him into the highest echelons of Budapest society. Three years later, a forty-three-year-old Max was rewarded with a hereditary title from the Austrian emperor Franz Joseph I for 'meritorious services in the financial field'. Max, a romantic, chose Margitta (then in Hungary, now in Romania) as the town that was to be associated with his title, traditionally the location of the family seat. Max's only connection to the place, however, was that the patron saint of the local church had the same name, Margit, as his wife. So the Neumanns became margittai Neumann (Neumann of Margitta) in Hungarian, and Max chose three marguerites (a type of daisy) for the insignia on the family's coat of arms. Many of the wealthy Jewish families ennobled during that period (more than 200 between 1900 and 1914) changed their names to more Germanic- or Hungarian-sounding ones to assimilate and often changed their faith too. Proud Max, though never particularly observant, did neither. Jancsi, who rather enjoyed the trappings of nobility when he was older, would adopt the Germanicized version of the name, first becoming Johann Neumann von Margitta while studying in Switzerland, then losing the place name to become simply 'von Neumann' in Germany.[13] After Max's death in 1928, his three boys converted to Catholicism for reasons similar to those of Heine.

In the same year that the Neumanns joined the European aristocracy, preparations were being made for Jancsi to start school. In much of Europe, a 'gymnasium' is a school that prepares students for further study at a university. Nearly all the Martians went to one of three elite fee-paying gymnasia in Budapest.

The foremost of these was the Minta or Model *gimnázium*,

Von Neumann doing mathematics, aged eleven, with his cousin Katalin (Lili) Alcsuti. *Courtesy of Marina von Neumann Whitman.*

founded in 1872 by Mór von Kármán, one of Hungary's leading education experts and, like Max, an ennobled Jew. The Minta was a test bed for von Kármán's educational theories, largely imported from Germany. Discipline and rigour were central, and education was based on problem-solving rather than rote-learning. 'At no time did we memorize rules from the book,' says von Kármán's son, Theodore, who attended the school. 'Instead we sought to develop them ourselves. In my case the Minta gave me a thorough grounding in inductive reasoning, that is, deriving general rules from specific examples – an approach that remained with me throughout my life.'[14] The younger von Kármán was to become the twentieth century's leading expert on aerodynamics and would shape the aircraft designs of both the German Luftwaffe (inadvertently) and the US Air Force.

The Minta's methods were successful enough to be widely copied by other schools including the older 'Fasori' Lutheran gymnasium, regarded as second only to the Minta itself. The Lutheran school was open to boys (there were scant educational opportunities for girls) of all faiths. Because the professional classes in Budapest were dominated by Jews, most of the students at the Lutheran school were in fact of Jewish descent.

The third option was a Real school (pronounced Re-Al, in the German fashion). These *reáliskola* provided a technical education, usually teaching no Greek and little Latin. 'The *reáliskola* was not at all inferior to the *gimnázium*, just different in scope and somewhat more practical than the "gentlemanly" *gimnázium*,' according to one historian and 'boasted extraordinary students in mathematics and the sciences'.[15] Among them were Fejér, Leo Szilard, who first conceived of the nuclear chain reaction that powers reactors and bombs, and Dennis Gabor, who won the Nobel Prize in physics in 1971 for inventing the hologram. One Real school in particular, located in District VI of Budapest, was considered to be on a par with the two gymnasia. From these three, Max chose the Lutheran gymnasium. The Minta's methods were too new-fangled to be trustworthy. The Real, on the other hand, lacked the classical education that he prized.

Some suggest these apparent genius factories were responsible for the great outpouring of Hungarian brilliance between 1880 and 1920. However, not all their ex-pupils concurred. Szilard, who attended the thoroughly modern and well-equipped Real school in District VI, found the maths classes 'intolerably boring' and, in an interview, called his teacher 'a complete idiot'.[16] Another of the Martians, Edward (Ede) Teller, joined the Minta in 1917, almost twenty years after von Kármán had left, and found his time there a trial. The maths classes 'set me back several years', he complained in his memoirs. 'Challenging students to explore ideas was not a common aim at the Minta.'[17]

Others believe the 'Hungarian phenomenon' was driven by two apparently contradictory elements of Hungarian society at the time: liberalism and feudalism. It was easier for Jews to rise to prominence in Austria-Hungary than in many of its less liberal European neighbours but the levers of power, in particular the civil service and military, were almost entirely in the hands of the Hungarian upper classes. Known derisively as the 'sandaled' nobility, this often-impoverished aristocracy was suspicious of the many non-Hungarians in their sprawling country whose loyalty to the old order was questionable. They allowed the new Jewish émigrés to prosper in professions including banking and medicine that they regarded as beneath them and granted the most successful – like Max – hereditary titles as a means of cementing their loyalty. Of the Martians, all were from Jewish backgrounds, all were wealthy, and two were titled.

Von Neumann himself attributed his generation's success to 'a coincidence of some cultural factors' that produced 'a feeling of extreme insecurity in the individuals, and the necessity to produce the unusual or face extinction'.[18] In other words, their recognition that the tolerant climate of Hungary might change overnight propelled some to preternatural efforts to succeed. Physics and mathematics were safe choices for Jews who wished to excel: an academic career could be pursued in many countries, and the subjects were viewed – in the early twentieth century, at least – as relatively harmless. Moreover, one could reasonably hope that good work in these fields would be fairly rewarded.[19] The truth of general relativity was established through experiment and was not contingent on whether the person who developed the theory was Jew or Gentile.

Whatever the relative contributions of schooling, upbringing and Hungarian society, in von Neumann's case everything propitiously aligned to produce a mathematical mind of rare ability. Jancsi started at the Lutheran in 1914, where he was about to prove he was no ordinary schoolboy. The foundations of mathematics were being shaken by the discovery of paradoxes that threatened to bring down the entire edifice. Some contended that hundred-year-old theorems that did not pass rigorous new standards of proof should be banished altogether. A battle for the soul of mathematics would soon follow. The very notion of truth was at stake. The seventeen-year-old von Neumann stepped in to put things right.

2

To Infinity and Beyond

A teenager tackles a crisis in mathematics

'Mathematics is the foundation of all exact knowledge of natural phenomena.'

David Hilbert, 1900

Von Neumann's unique talents were spotted as soon as he started school. He attracted the attention of the Lutheran school's legendary maths teacher, László Rátz, who is so venerated in Hungary that a street in Budapest is named after him. Rátz quickly concluded that von Neumann would benefit from a mathematical education far more advanced than he could provide. He arranged to see Max and offered to organize extracurricular tuition for Jancsi at the University of Budapest. Rátz promised his pupil would continue to receive the full benefits of the Lutheran's classical education by attending all classes (including, somewhat redundantly, maths). Max, well aware of his son's mathematical talents, agreed. Rátz refused to accept any money for his services. The privilege of teaching Jancsi, he felt, was sufficient.

The young von Neumann made an instant impact on his new tutors. His first mentor, Gábor Szegő, who would later lead Stanford University's maths department,[1] was moved to tears after their first meeting. The most influential of von Neumann's tutors was Lipót Fejér, a seminal figure in Hungarian mathematics who drew many of the country's most talented stars, including, earlier, Szegő himself. 'There was hardly an intelligent, let alone a gifted, student who could exempt himself from the magic of his lectures,' says fellow Hungarian mathematician George Pólya. 'They could not resist imitating his stress patterns and gestures, such was his personal impact upon them.'[2] Fejér's interest in his young charges went well beyond the call of duty. 'What does little Johnny Neumann do?' he would write to Szegő years later. 'Please let me know what impact you notice so far of his Berlin stay.'[3] Szegő was by then teaching at the University of Berlin, where 'little Johnny' was ostensibly boning up on undergraduate chemistry. (He was mostly creaming off all he could from the best at the university's prestigious mathematics department.)

Fejér and Michael (Mihály) Fekete, another of Fejér's former students, took on the brunt of von Neumann's education during his teenage years. All three of these tutors – Szegő, Fejér and Fekete – shared an interest in orthogonal polynomials, so it was natural that they were the subject of von Neumann's first paper. Orthogonal polynomials are sets of independent mathematical functions that can be added together to make any other. The complicated heaving and swaying of a ship at sea, for example, can be broken down into a simpler sum of these functions (a process known as harmonic analysis) and plugged into a computer to simulate the vessel's motion. This facility for making messy real-world data more manageable is why such polynomials are often used in physics and engineering.

For mathematicians, a key characteristic of polynomials is their 'zeroes'; that is, the question of where they cross the x-axis on a graph. Von Neumann's first paper,[4] written with Fekete, investigated the zeroes of Chebyshev polynomials, discovered as a result of a Russian mathematician's obsession with the problem of how to most efficiently turn the up and down motion of a steam engine piston into the circular motion of a wheel.[5]

This was von Neumann's introduction to the norms and conventions of academic maths. He was just seventeen when the completed manuscript was sent away for publication. Mathematicians have their own style much as novelists do. Von Neumann's voice appears here for the first time, more or less fully formed. 'Johnny's unique gift as a mathematician was to transform problems in all areas of mathematics into problems of logic,' says Freeman Dyson.

> He was able to see intuitively the logical essence of problems and then to use the simple rules of logic to solve the problems. His first paper is

> a fine example of his style of thinking. A theorem which appears to belong to geometry, restricting the possible positions of points where some function of a complex variable is equal to zero, is transformed into a statement of pure logic. All the geometrical complications disappear, and the proof of the theorem becomes short and easy.[6]

While von Neumann would never again mention this treatise in any future work, Fekete, inspired by the prodigy, would devote most of his remaining career to the subject.

Hungary, by this time, had fought in and lost a world war. But Budapest was never close to the front, and life for the wealthy denizens of Vaczi Boulevard had continued largely as before. Their lives were temporarily disrupted by a coup and the establishment in 1919 of Europe's first communist government (after Russia), led by Béla Kun, a non-practising Hungarian Jew who had become a convert to the revolutionary cause as a Russian prisoner of war. Kun was officially the new administration's foreign minister, but his popularity ensured that he held the reins of power. 'My personal influence in the Revolutionary Governing Council is such that the dictatorship of the proletariat is firmly established,' he told Lenin, 'since the masses are backing me.'[7]

As armed party enforcers in leather jackets known as the 'Lenin Boys' stalked the streets of Budapest, the Neumann family packed their bags and left for a holiday home on the Adriatic – but not before Max had secured their Budapest dwelling from being requisitioned by the new government. 'Under the guiding principle of equal facilities to all, the big apartments were broken up,' said Nicholas, who was seven at the time. But the party members charged with the task were soon convinced to forget the matter. 'On the piano under a weight, my father put a bundle of British pound notes, I don't know how much. The Communist official with the red armband promptly went there, took it, and the committee left and we remained in the apartment.'[8]

After Kun waged an ill-conceived war to re-establish Hungary's pre-First World War borders, the Romanian army marched into Budapest to topple his 133-day-old Hungarian Soviet Republic. Kun fled, eventually to live in exile in Russia, where he was executed in 1937 after being branded a Trotskyist and enemy of the people. The chaos

The von Neumanns visiting an army artillery post ca.1915. John is sitting on the gun barrel. The three children on the gun carriage are (from top left to bottom right) brother Michael, cousin Lili, and brother Nicholas (in a dress). *Courtesy of Marina von Neumann Whitman.*

of the Kun regime would remain etched on von Neumann's mind. 'My opinions have been violently opposed to Marxism ever since I remember,' he told a congressional confirmation hearing in 1955, after he was nominated to the Atomic Energy Commission, 'and quite in particular since I had about a three-month taste of it in Hungary in 1919.'[9]

Max had meanwhile left Budapest for Vienna to contact supporters of Admiral Miklós Horthy, a war hero who was to assume control of the counter-revolutionary forces amassing against Kun. In the chaos that followed the downfall of Kun's government, Horthy's forces ranged through Hungary exacting revenge on anyone they felt had sympathized with the communists. Jews had occupied prominent positions in Kun's short-lived government and soon became the focus of their ire. The Lenin Boys and their ilk had despatched 500 souls during the Red Terror. In the ensuing White Terror, Horthy's officers would kill around 5,000. Rape, torture and public hangings were commonplace. Bodies were mutilated as a warning to others. Horthy mildly rebuked one of his most brutal lieutenants

> for the many Jewish corpses found in the various parts of the country . . . This, [Horthy] emphasized, gave the foreign press extra ammunitions against us . . . in vain, I tried to convince him that the liberal papers would be against us anyway, and it did not matter that we killed only one Jew or we killed them all.[10]

The von Neumanns were spared by Horthy's forces and, miraculously, von Neumann's schooling continued through the upheaval more or less undisturbed. Two of his schoolmates during these years were to remain his friends for life. Eugene (Jenő) Wigner was a year ahead of him at the Lutheran. William (Vilmos) Fellner, who was to become an eminent economist and a professor at Yale University, was in the year below. They remembered a boy acutely aware of his intellectual superiority, who was neither particularly popular nor disliked. Displaying a sensitivity towards the feelings of others that is not always found in those with remarkable brains, von Neumann took care not to be overbearing yet could not but help stand apart. 'Whenever I talked with von Neumann,' Wigner said of his friend, 'I always had the impression that only he was fully awake, that I was halfway in a dream.'[11] Jancsi quickly outgrew children's games, the two friends recall, and took to observing those around him with the disquieting, detached manner of an anthropologist.

Modernism was by now sweeping through mathematics as it was through art, music and literature. As von Neumann was revising for his final exams at the gymnasium in 1921, Piet Mondrian, dissatisfied with the limitations of realism, was painting *Tableau I*, his first grid-based abstract work with colourful arrangements of red, blue and yellow squares. The poet and critic Guillaume Apollinaire summed up the motivations of the radicals: 'Real resemblance no longer has any importance, since everything is sacrificed by the artist to truth, to the necessities of a higher nature whose existence he assumes, but does not lay bare.'[12]

During von Neumann's school years, this impulse to look beyond the surface of things had spread decisively to mathematics. Assumptions accepted for a thousand years or more were being probed – and found wanting. The ensuing foundational crisis was no genteel disagreement between beard-tugging worthies but a struggle for the heart and soul of mathematics, attracting some of the best minds of the day, with consequences that would echo down to the present. The objectives and status of mathematics would be changed for ever. Once regarded as a divine fount of truth, mathematics would be shown to be a thoroughly human endeavour; imperfect and destined to remain so.

Von Neumann would learn of the upheaval in maths as a teenager, and his efforts to resolve the crisis, in the form of a brilliant series of papers, sealed his reputation as a genius of the highest order. When he later turned from pure mathematics to real-world problems, his contributions to the foundational crisis prepared him, quite unexpectedly, for an intellectual leap that would bring the modern computer into being. A heated debate on the limits of mathematics would, in time, spawn Apple, IBM and Microsoft.

The roots of the foundational crisis lay in the discovery of flaws in Euclid's *Elements*, the standard textbook in geometry for centuries. Euclid had devised five statements that he assumed to be self-evident (his axioms or postulates). By building on these through a series of logical steps, he proved a number of more complex statements (theorems), including Pythagoras' theorem: that the square of the longest side of a right-angled triangle is equal to the sum of the squares of the other two sides. This 'axiomatic method' was the cornerstone of mathematics, and the planets were assumed to wheel through the sort of three-dimensional space the *Elements* described. At the start of the nineteenth century, only one true geometry was thought possible. 'Euclidean geometry was a repository of truths about the world that were as certain as any knowledge could be,' says historian Jeremy Gray. 'It was also the space of Newtonian physics. It was the geometry dinned into one at school. If it failed, what sort of useful knowledge was possible at all?'[13]

That first step towards shattering that orthodoxy was taken in the 1830s by János Bolyai, another Hungarian mathematical prodigy, and Nicolai Lobachevsky, a Russian. Both independently developed geometries in which the last of Euclid's five statements – his 'parallel postulate' – was not true. Compared with the other postulates, the fifth stood out. The second postulate, for example, says that any line segment may be extended indefinitely. That is difficult for even the most querulous to argue with. The fifth, on the other hand, states that if two lines are drawn which intersect a third in such a way that the sum of the inner angles on one side (labelled *a* and *b* in the diagram below) is less than two right angles (i.e. 180°), then the two lines inevitably must intersect each other on that side if extended far enough. If, on the other hand, a and b *do* add up to 180°, the two lines never meet so are said to be parallel.

To mathematicians, that looks less like a postulate and more like a theorem in need of proving. And for 2,000 years, many tried – and failed – to do just that. When Bolyai's father, himself a geometer,

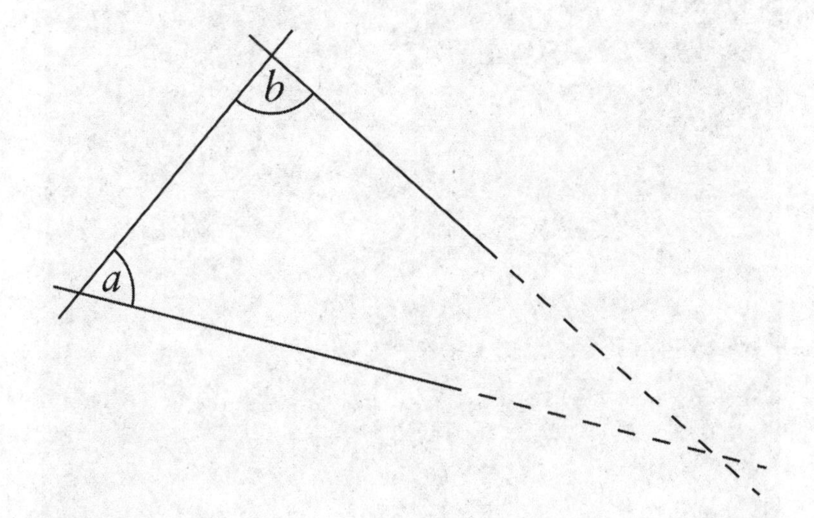

Euclid's parallel postulate.

heard about his intention to work on the fifth postulate, he urged him to desist: 'Learn from my example: I wanted to know about parallels, I remain ignorant, this has taken all the flowers of my life and all my time from me.'[14] When János showed him his work, however, the elder Bolyai's worries were somewhat allayed.

What Bolyai and Lobachevsky had discovered is now known as hyperbolic geometry. Whereas the sort of surface for which all five of Euclid's postulates hold is flat, like a sheet of paper, a hyperbolic surface curves away from itself everywhere like a saddle. Picture one of those well-known stackable potato-crisps or the wrinkled fruiting body of a wood ear mushroom. On these sorts of surfaces, many of the familiar rules of school geometry no longer hold: the three angles of a triangle, for instance, add up to less than 180°. 'Out of nothing,' wrote Bolyai to his father, 'I have created a strange new universe.'

The next leap forward in geometry was taken by the German mathematician Bernhard Riemann in the 1850s, about twenty years after Bolyai and Lobachevsky. Riemann's doctoral thesis, now recognized as

The hyperbolic surfaces of a wood ear mushroom.

one of the greatest ever produced in mathematics, was of 'a gloriously fertile originality,' said Carl Friedrich Gauss, the most famous mathematician of the age. While Bolyai and Lobachevsky pictured planes that curved through space, Riemann's surfaces often twisted and contorted in ways that could barely be imagined at all. His mathematics could describe space with any number of dimensions – hyperspace – just as easily as the three familiar spatial ones. More than half a century later, Riemann's geometry would prove 'admirably appropriate' for describing the warped four-dimensional spacetime of Einstein's general relativity.

By the late nineteenth century, many other assumptions and proofs in Euclid's work were being questioned. Some argued that it was time to start from scratch by rebuilding geometry on new foundations. The task was taken up by David Hilbert, who was to become the most influential mathematician of the early twentieth century. His resulting book, *Grundlagen der Geometrie* (*The Foundations of Geometry*), published in 1899, is now recognized in its clarity of reasoning as the rightful heir of Euclid's *Elements*. A mathematical best-seller, its impact was immediate.

Hilbert's aim was no less than to distil, from the works of his predecessors, trustworthy ways to reason about *any* elementary geometry. To avoid relying on the intuition of his readers, the familiar terms of school geometry (points, lines, planes, etc.) were emptied of their meaning. In his book these words were simply convenient labels for mathematical objects that would be rigorously defined by the mathematical relationships between them.[15] 'One must be able to say at all times – instead of points, straight lines, and planes – tables, chairs and beer mugs,' Hilbert had explained years earlier.[16] The advantage of this incredibly abstract approach to geometry was that his findings would be valid for *any* set of objects as long as they obeyed the rules he had carefully concocted.

Hilbert defined his axioms with a rigour that went well beyond Euclid's. His improved axiomatic method would come to define the way mathematics was done in the twentieth century. Hilbert's book on geometry's foundations cemented his reputation as a great mathematician. Not yet forty and secure in his chair at the University of Göttingen, he set about proving himself to be a master administrator by attracting funding and talent. By 1920, there was no mathematics department in the world to rival that of Göttingen.

In his position as the discipline's foremost spokesperson, Hilbert now demanded that all of mathematics – and indeed the sciences – be made as bulletproof as his new geometry. In 1880, the famous physiologist Emil du Bois-Reymond had declared there were some questions (he called them 'world riddles'), such as the ultimate nature of matter and force, that science could never answer. '*Ignoramus et ignorabimus*,' as he put it: 'we do not know and will not know'.

Hilbert was having none of it. In 1900, he sounded his opposition to du Bois-Reymond's pessimism, denying that there were any such limits to knowledge. Every question had a definite answer, he argued, even if what that answer showed was that answering the original question was impossible. At the International Congress of Mathematicians in Paris that year, he posed twenty-three problems that would shape mathematics in the twentieth century. 'For us there is no *ignorabimus*, and in my opinion none whatever in natural science,' Hilbert was to thunder even thirty years later. 'In opposition to the foolish *ignorabimus* our slogan shall be: *Wir müssen wissen – wir werden wissen*.' We must know – we will know. Many rallied to Hilbert's side, eager to make mathematics (and so, it followed by Hilbert's logic, the sciences) unassailable. But his project ran into the sand almost as soon as it was conceived.

In 1901, the British philosopher and logician Bertrand Russell found a paradox at the heart of set theory, a branch of mathematics pioneered by Georg Cantor a quarter of a century earlier. Cantor, a brilliant, deeply religious Russian-born German Protestant, was the first mathematician to see that there are a multitude of different infinities, and some infinities are demonstrably larger than others. To the greatest of his infinities, the 'Absolute', Cantor assigned the Greek capital omega, Ω. Only in the mind of God, he said, could Ω be truly contemplated in all its glory. Realizing the controversial nature of the findings, he called his new infinities 'transfinite numbers' in order to distinguish them from the old concept of infinity. His insights, he said, came directly from God.

Not everyone agreed. 'God made the natural numbers; all else is the work of man,' growled Leopold Kronecker, a contemporary grandee of German mathematics who found Cantor's juggling with infinities suspicious and distasteful. He called Cantor a 'charlatan' and 'corrupter of youth' and squashed his hopes of moving from Halle University to a chair at the much more prestigious University of Berlin. Cantor was ill-equipped emotionally to deal with the vitriolic response to his transfinite numbers. Kronecker's attacks precipitated a bout of depression and the first of many visits to a sanatorium.

By the time Russell began his work, set theory was considered rather more brilliant than suspicious. Mathematics does not ultimately deal with finite groups of numbers. If, for example, a mathematician wants to prove something about prime numbers then, usually, the goal is to find a theorem that applies equally well to *all* prime numbers – of which there are infinitely many. Mathematicians had embraced Cantor's theory as a powerful tool for manipulating and proving theorems about such sets of infinite size.

Russell's paradox, however, threatened to deal a far more serious blow to set theory than earlier ideological objections. The problem was this: consider a set of objects – all possible types of cheesecake, say. This set may include any number of different cheesecakes (New York cheesecake, German *Käsekuchen*, lemon ricotta, etc.) but, because a set is not literally a cheesecake, the set of all cheesecakes is not a member of itself. The set of all things that are *not* cheesecakes, on the other hand, *is* a member of itself.

But what, Russell wondered, about the set of *all* sets that are not members of themselves. If this is *not* a member of itself, then, by definition, it should be (because its members do not include itself). Conversely, if it *is* a member of itself, then it should not be (because it does). This was Russell's paradox in a nutshell. His analysis of the paradox revealed it to be similar in form to several others, including the liar's paradox ('this statement is a lie'). 'It seemed unworthy of a grown man to spend time on such trivialities,' he complained, desperate for a solution, 'but what was I to do?'[17]

Russell had begun a huge effort to describe precisely the logical basis of all of mathematics, but was plunged into despair by his discovery. Unable to progress with his work, he spent the next several years trying, without success, to resolve the contradiction he had found. 'Every morning I would sit down before a blank sheet of paper,' he said. 'Throughout the day, with a brief interval for lunch, I would stare at the blank sheet. Often when evening came it was still empty . . . it seemed quite likely that the whole rest of my life might be consumed in looking at that blank sheet of paper.'[18]

Russell's paradox and others like it threatened to kick away a cornerstone of mathematics and with it Hilbert's programme of re-establishing mathematics on more rigorous grounds. Alarmed, Hilbert called for mathematicians to resolve the crisis Russell's discovery had triggered. 'No one', he vowed, 'will drive us out of this paradise that Cantor has created for us.'[19]

Not all mathematicians were as resolved to save Cantor as Hilbert was. One group, the 'intuitionists', led by the pugilistic and highly strung young Dutch mathematician L. E. J. 'Bertus' Brouwer, argued Russell's paradox showed that mathematics was bumping up against the limits of the human mind. Brouwer was wary of transfinite numbers. There was no reason to believe, he argued, that the rules of logic could be applied to everything in mathematics and, in particular, not to Cantor's dubious infinite sets. The law of the excluded middle, for example, states that either a proposition or its negation is true. So the sentence 'I am a dog' is either true or false but cannot be both. Brouwer contended that to prove this satisfactorily for a set, each of its members must be inspected to establish whether the proposition did or did not hold. To do this for the members of an infinitely large set is, of course, impossible. Playing fast and loose with infinite sets, Brouwer claimed, had led to paradoxes of the sort that so troubled Russell.

In Göttingen, Hilbert reacted angrily. Whereas once he had supported Brouwer's application for a chair at the University of Amsterdam, he now campaigned for him to be removed from the board of editors of *Mathematische Annalen*, one of the field's most prestigious journals. Einstein, one of the editors, dismissed the feud as a completely overblown *Froschmäusekrieg* (literally a war of frogs and mice, a German phrase describing a bitter but unimportant altercation). For Hilbert, however, this was not merely some trivial spat about mathematical niceties. There was much more at stake. 'If mathematical thinking is defective,' he asked, 'where are we to find truth and certitude?'

*

For any ambitious young mathematician determined to prove his mettle, the idea of saving mathematics from itself was irresistible. Despite his tender years, von Neumann was well equipped for the task. He had lectured Wigner enthusiastically on the delights of set theory on long weekend walks when he was eleven years old. In 1921, the brash seventeen-year-old came to Hilbert's aid by trying to resolve the crisis that had stumped many of the cleverest mathematicians in the world. His first contribution secured numbers themselves from the sort of paradoxes that Russell had found.

The notion of number in Cantor's theory is related to two essential characteristics of sets: cardinality and ordinality. Cardinality is a measure of the size of a set: a set with three members, for example, is of cardinality 3. Ordinality, on the other hand, designates how a set is ordered. This is related to the ordinal numbers (1st, 2nd, 3rd ...) which specify the position of elements in a set. Cardinality was formally defined as the set of all sets of equal cardinality. That is, a set containing five things is of the same cardinality (five!) as all other sets containing five things. Ordinality was defined in a similar way. This now looked dangerously circular. Since both concepts are necessary if mathematicians are to be able to manipulate sets and prove theorems, von Neumann wanted to excise all talk of 'sets of all sets' from their definitions – a first step towards saving 'Cantor's paradise'.

Von Neumann's paper exudes the sort of confidence that might be expected of an established master rather than a schoolboy. His first paragraph is a single sentence: 'The purpose of this work is to make the idea of Cantor's ordinal numbers unambiguous and concrete.'[20] He does so in seventeen carefully argued logical steps, described in a total of ten pages. In plain but less precise language, von Neumann begins by defining the ordinal '1st' to be the empty set. Then he defines a recursive relationship, such that the next highest ordinal is the set of all smaller ordinals. '2nd' is therefore the set containing only '1st', which is the empty set. '3rd' is the set containing '2nd' and '1st', which is both the set containing the empty set and the empty set itself. '4th' is the set containing the preceding ordinals: '3rd', '2nd' and '1st' and so on. The process is a bit like building successively taller towers adjacent to each other with Lego bricks. '1st' might be a single red brick; '2nd' a red brick, and a two-block tower next door. The work continues until you reach the ordinal of your choice.

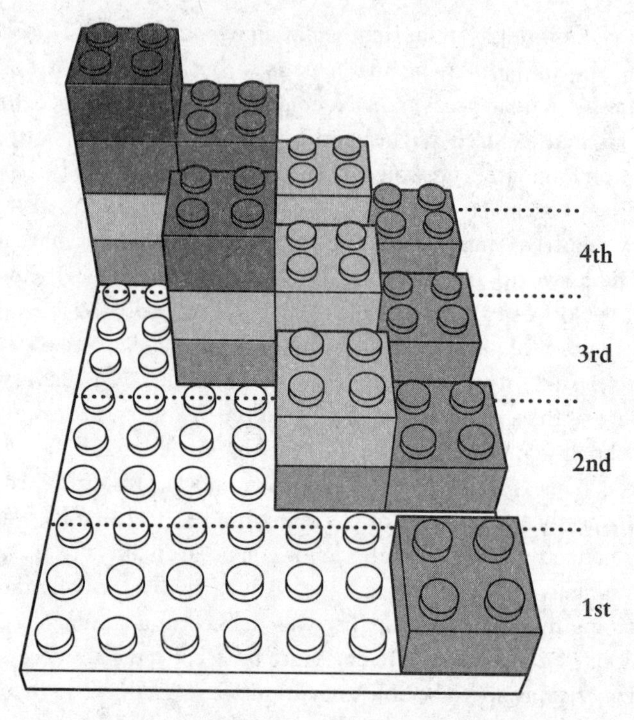

Von Neumann's approach to ordinals expressed in Lego.

Cardinal numbers can then be defined by putting them in a 'one-to-one correspondence' with ordinals; that is, pairing them off with each other: 0 with 1st (the empty set), 1 with 2nd (which contains one element), 2 with 3rd (which contains two elements) and so on. Von Neumann's definition appears so deceptively straightforward that a non-mathematician may wonder why he took ten pages to lay out his thinking. The answer is that many simple-seeming ideas were turning out to contain troubling contradictions. Von Neumann, following Hilbert's rigorous axiomatic method, wanted to make sure his did not. A testament to his success is that nearly a hundred years after his paper was published, von Neumann's definition of cardinal and ordinal numbers remains the standard one in mathematics today.

Still, the paradoxes remained, casting a shadow over the trustworthiness of set theory. Von Neumann, who was about to graduate from high school, was eager to help, but there was an obstacle in the shape of his father to overcome first. Max, concerned that his prodigy was taking an overly strong interest in mathematics, approached Theodore von Kármán, around twenty years Jancsi's senior and already a renowned aerospace engineer, to dissuade him from pursuing the subject at university. 'Mathematics', he explained to von Kármán, 'does not make money.'[21] Dutifully, von Kármán arrived at Vaczi Boulevard to probe the young von Neumann's interests. 'I talked with the boy,' said von Kármán of the encounter. 'He was spectacular. At seventeen he was already involved in studying on his own the different concepts of infinity, which is one of the deepest problems of abstract mathematics . . . I thought it would be a shame to influence him away from his natural bent.'

Max was adamant, however, so von Kármán helped to forge a compromise. Jancsi would work towards a degree and a doctorate in two different subjects at the same time. The chemical industry was in its heyday, so he would catch up on chemistry at the University of Berlin with a view to applying, after two years, for a place to study chemical engineering at the Swiss Federal Institute of Technology (ETH) in Zurich. He would also register as a doctoral student in mathematics at the University of Budapest.[22]

Von Neumann sailed through his final exams at the Lutheran school, as expected. With the exception of three areas – physical education, music and handwriting – his grades had rarely dipped below 'excellent'. He was not an angel – conduct was usually marked as merely 'good' – but he must have been unimaginably bored in most lessons.

With his school years behind him, von Neumann took the train to Berlin with his father in September 1921 to begin the arduous programme of study that had been agreed. A passenger sharing their carriage, having learned a little about his interests, looked to engage the youngster in some friendly chit-chat: 'I suppose you are coming to Berlin to learn mathematics.' 'No,' von Neumann replied, 'I already know mathematics. I am coming to learn chemistry.'

So began the frenetic, peripatetic activity that would characterize much of the rest of his life. Von Neumann shuttled busily between three cities for the next five years. In September 1923, after catching up on basic chemistry at Berlin, he sat the ETH entrance exam and passed with flying colours. During the next three years he doggedly pursued chemical engineering, racking up a record bill for broken glassware at the ETH that would not be topped for some time. His heart, however, was elsewhere. Whether it was in Berlin, Zurich or Budapest, von Neumann was finding mathematicians to talk to. In Berlin, he became the protégé of Erhard Schmidt, who had studied under Hilbert twenty years earlier. Later in Zurich, he turned to Hermann Weyl, regarded as the best of Hilbert's former students, who welcomed him with open arms and would join him in Princeton a decade later.

The result of all von Neumann's to-ing and fro-ing and the tutelage of mathematicians twice his age arrived on the desk of Abraham Fraenkel, one of the world's leading experts in set theory, sometime between 1922 and 1923. Aged nineteen, von Neumann had written a first draft of his doctoral thesis. Fraenkel would later recall receiving 'a long manuscript of an author unknown to me, Johannes von Neumann, with the title "Die Axiomatisierung der Mengenlehre" ("The Axiomatization of Set Theory") . . . I don't maintain that I understood everything, but enough to see that this was outstanding work and to recognize *ex ungue leonem*.'[23] 'To know the lion by his claw' – the Latin words uttered by Johann Bernoulli 200 years earlier, on recognizing by its brilliance an anonymous work by Isaac Newton.

Fraenkel asked von Neumann to make his theory more comprehensible to mere mortals. The revised manuscript, with the 'The' changed to a more modest 'An', was published in 1925.[24] Over the next three years, von Neumann expanded the paper and published the longer version with the 'An' changed back to a 'The'.[25] In it, to Hilbert's delight, he placed set theory on solid ground and provided a simple way out of Russell's paradox.

Russell's own efforts to tackle his paradox was his 'theory of types', which he laid out most definitively in the *Principia Mathematica*. Published in three weighty volumes between 1910 and 1913, the *Principia* was an effort to describe axioms and rules from which all of mathematics could be derived. After 379 pages, Russell and his co-author Alfred North Whitehead were able to prove that 1+1=2 ('The above proposition is occasionally useful,' read the wry comment accompanying their proof). Type theory tries to avoid circular statements by organizing them into collections ('types') and assigning to these collections a strict ordering. Crucially, statements that fully define the membership of a set take precedence over those that ask about the properties of the set. Asking whether 'the set of all sets that are not members of themselves' is a member of itself is therefore redundant: the set's membership is defined first to avoid such contradictions. Russell's type theory was quite ponderous, however, and by placing strict limits on what could and could not be said, it also threatened to limit the scope of mathematics.

Von Neumann's approach, by comparison, is beautifully simple. He lists all his axioms on a single page. 'This is sufficient to build up practically all of the naive set theory and therewith all of modern mathematics and constitutes, to this day, one of the best foundations for set-theoretical mathematics,' wrote the mathematician Stanisław Ulam decades later. 'The conciseness of the system of axioms and the formal character of the reasoning employed seem to realize Hilbert's goal of treating mathematics as a finite game,' continues Ulam, who would become one of von Neumann's closest friends. 'Here one can divine the germ of von Neumann's future interest in computing machines and the "mechanization" of proofs.'

The paper resolved Russell's paradox by distinguishing between two distinct sorts of collections. He called them *I. Dingen* and *II. Dingen*: 'one things' and 'two things'. Mathematicians now tend to call them, respectively, 'sets' and 'classes'. Von Neumann rigorously defines a class as a collection of sets that share a property. In his theory it is no longer possible to speak meaningfully of either a 'set of all sets' or a 'class of all classes'; only a 'class of all sets'. Von Neumann's formulation elegantly avoids the contradictions of Russell's paradox without all the restrictions of type theory. There is no 'set of all sets that are not members of themselves' but there is a 'class of all sets that are not members of themselves'. Crucially, this class is not a member of itself because it is not a set (it's a class!).

Von Neumann's paper confirmed that he was no flash in the pan. By the time it was published in 1925, he had added another town to his list of regular haunts. To the annoyance of some of the old guard in Göttingen, he had become a favourite of Hilbert. The pair would go for walks in Hilbert's garden or lock themselves up together in his study to discuss the foundations of mathematics and the new quantum theory, which was flowering untidily around them. The following year, von Neumann both graduated from his chemical engineering degree at the ETH and passed his doctoral exam with aplomb. He was twenty-two. Hilbert, one of the examiners, is alleged to have asked just one question. 'In all my years I have never seen such beautiful evening clothes: pray, who is the candidate's tailor?'[26]

Meanwhile, Hilbert, buoyed by how well his programme was proceeding, set out exactly what was required to ensure that mathematics was secure once and for all. In 1928, he challenged his followers to prove that mathematics is complete, consistent and decidable. By complete, Hilbert meant that all true mathematical theorems and statements can be proved from a finite set of axioms. By consistent, he was demanding a proof that the axioms would not lead to any contradictions. The third of Hilbert's demands, that mathematics should be decidable, became widely known as the *Entscheidungsproblem*: is there a step-by-step procedure (an algorithm) that can be used to show whether or not any particular mathematical statement can be proved? Mathematics would only be truly safe, said Hilbert, when, as he expected, all three of his demands were met.

Hilbert's dreams of a perfect mathematics were soon to be crushed. Within a decade, some of the brightest minds in mathematics answered his call. They would show that mathematics was neither complete nor consistent nor decidable. Shortly after Hilbert's death in 1943, however, his failed programme would bear unexpected fruit. Hilbert had driven mathematicians to think extraordinarily systematically about the nature of problems that were or were not solvable through step-by-step, mechanical procedures. Through von Neumann, that abstruse pursuit would help to birth a truly revolutionary machine: the modern computer.

3

The Quantum Evangelist

How God plays dice

'If only I knew more mathematics!'

Erwin Schrödinger, 1925

After his doctoral examination, von Neumann quickly secured a grant from the Rockefeller Foundation and headed to Hilbert's Göttingen, the centre of the mathematical world. Also in Göttingen at that time there was another boy wonder, the twenty-three-year-old Werner Heisenberg, who was laying the groundwork for a successful – but bewildering – new science of the atom and its constituents, which was soon christened 'quantum mechanics'. The theory would explain many odd experimental results from preceding decades but threatened to overturn cherished ideas scientists had held about the nature of reality for hundreds of years. Quantum mechanics opened a rift between cause and effect, banishing the Newtonian clockwork universe, where tick had reliably followed tock.

Beginning in 1900, when German physicist Max Planck reluctantly introduced the radical idea that energy might be absorbed or emitted in lumps, or 'quanta', discoveries that had seemed at first trivial would challenge, then revolutionize, physics just as Russell's paradoxes were shaking the foundations of mathematics. Building upon Planck's idea, Einstein would in 1905 theorize that light itself might be composed of a stream of particles, the first hint that quantum entities had both wave-like and particle-like properties.

While von Neumann was still a schoolchild, Danish physicist Niels Bohr was busily cobbling together a new model of the atom that awkwardly melded Newtonian physics with the 'quanta' of Planck and Einstein. In Bohr's quantum atom of 1913, electrons could occupy only certain special orbits and jumped from one orbit to another by absorbing a chunk of energy *exactly* equal to the difference in energy between them.

Brilliant though it was, Bohr's model was a jury-rigged affair that raised as many questions as it answered. What held electrons in their 'special' orbits? How did they hop from one to another in an instant? 'The more successes the quantum theory enjoys, the more stupid it looks,' said Einstein, who realized early on that the shotgun wedding of classical and quantum concepts could not last.[1] Physicists soon wanted an amicable divorce.

In 1925, Heisenberg formulated the first rigorous approach to quantum theory, now known as 'matrix mechanics'. His jauntily titled *Quantum Theoretical Re-interpretation of Kinematic and Mechanical Relations* landed like a bomb at the end of that summer.[2] But as von Neumann arrived in Göttingen in 1926, Erwin Schrödinger, a professor at the University of Zurich, proposed a completely different approach to quantum mechanics based on waves. Despite bearing no resemblance to Heisenberg's matrix mechanics, Schrödinger's 'wave mechanics' worked just as well. Could two such wildly different looking theories be describing the same quantum reality? What followed was five of the most remarkable years in the history of science, during which a mechanics to describe the quantum world would be forged, much of it at Göttingen.

Having already made a name for himself by tackling some of the most vexing questions in mathematics, von Neumann now turned to this question, one of the biggest puzzles in contemporary physics. He would eventually show decisively that at the deepest level Heisenberg's and Schrödinger's theories were one and the same. Upon that insight, he would build the first rigorous framework for the new science, influencing generations and bringing clarity to the hunt for its meaning.

The core ideas in Heisenberg's revolutionary paper were assembled during a two-week stay in June 1925 at Heligoland, a sparsely inhabited rock shaped like a wizard's hat that lies some 30 miles north of the German coast. Puffed up by a severe attack of hay fever, the outdoorsman had gone to take the pollen-free North Sea air, hiking and swimming and hoping for a breakthrough that would make sense of the puzzles thrown up by Bohr's work. Heisenberg wanted a mathematical framework that would account for the things that scientists could actually see in the laboratory: chiefly, the frequencies and relative intensities of 'spectral lines'. Excite atoms, by vaporising a sliver of material in a flame or passing a current through a gas, and they will emit radiation. The bright colours of neon lights and the sickly yellow hue of a sodium vapour lamp are a result of the excited atoms within producing intense light at characteristic wavelengths. By the early twentieth century, that each element produced a unique set of spectral lines was well known.

Bohr had proposed these sharp spikes in the spectrum of radiated light were caused by excited electrons tumbling back to the ground state of an atom, in the process emitting light waves with energy equal to the difference between the higher and lower orbits. Heisenberg accepted this but he rejected the physical implications of Bohr's model – electrons had never been seen spinning in orbits around an atom's nucleus (nor would they be). Heisenberg instead stuck to the observed facts. He showed the frequencies of atomic emission lines could be represented conveniently in an array, with rows and columns representing, respectively, the initial and final energy levels of electrons producing them. When written like this, the frequency of radiation emitted by an electron falling from energy level 4 to 2, for example, would be found at row 4, column 2 of his array. But since electron transitions between different energy levels appeared to take place more or less instantly,[3] there was no way to know whether an electron had jumped to its final state directly or passed through an intermediate state on the way.[4] According to the laws of probability, the chance of two transitions occurring one after the other is equal to their individual probabilities multiplied together.[5] To find the overall probability of all possible transitions easily, Heisenberg arranged the individual transition probabilities in an array too, and multiplied rows and columns together.[6] When he did so, he discovered a strange property of his arrays: multiplying one array, A, by another, B often gave a different answer from multiplying B by A.[7] This troubled him because he knew ordinary numbers do not behave in this way. As every schoolchild learns, multiplying 3 by 7 gives the same answer as

$$f_{m,n} \left| \begin{array}{cccccc} f_{1,1} & f_{1,2} & f_{1,3} & f_{1,4} & f_{1,5} & \cdots\cdots \\ f_{2,1} & f_{2,2} & f_{2,3} & f_{2,4} & f_{2,5} & \cdots\cdots \\ f_{3,1} & f_{3,2} & f_{3,3} & f_{3,4} & f_{3,5} & \cdots\cdots \\ f_{4,1} & f_{4,2} & f_{4,3} & f_{4,4} & f_{4,5} & \cdots\cdots \\ f_{5,1} & f_{5,2} & f_{5,3} & f_{5,4} & f_{5,5} & \cdots\cdots \\ \vdots & \vdots & \vdots & \vdots & \vdots & \ddots \end{array} \right.$$

The frequencies of atomic emission lines can be written in arrays.

multiplying 7 by 3. Mathematicians say multiplication is commutative because the two numbers can be multiplied in either order: A×B = B×A. But this was not the case with Heisenberg's arrays. They did not commute.

Heisenberg returned to Göttingen, where he was an assistant to the theorist Max Born. He showed what he described as his 'crazy paper'[8] to Born, who encouraged him to publish, writing to Einstein that it was 'rather mystifying' but also 'true and profound'.[9] Only after the paper was in print did Born remember that he had been taught about similar arrays years earlier. Named 'matrices' by the English mathematician James Sylvester in 1850, their properties were elucidated by Sylvester's friend and collaborator Arthur Cayley, though Chinese mathematicians made use of them some 2,000 years earlier. Born had used them in a paper on relativity in 1909 and, crucially, as he now recalled, matrix multiplication is not commutative. Heisenberg had rediscovered a type of mathematics with ancient roots. (The basics of matrix algebra, so unfamiliar to him, are now taught to high-school children.)

Inspired by Heisenberg's work on transition probabilities, Born intuited a formula[10] connecting the position of a particle with its momentum, showing that the two did not commute either. Multiplying position by momentum or, conversely, momentum by position gives slightly different results. The difference (less than a trillionth of a trillionth of a billionth of 1 joule-second) is far too small to be noticed in everyday life but is large enough to be significant at the atomic scale. Pondering the physical meaning of noncommutativity led Heisenberg in 1927 to an extraordinary new law of nature, which stated that the position and momentum of a particle cannot both ever have exact values at the same time. And if it is impossible to know at any moment both the location and velocity of a particle exactly then one cannot, as physicists had long assumed, predict where it will be next. Heisenberg's insight was to become known as his uncertainty principle.[11]

Schrödinger's formulation of quantum mechanics, which appeared shortly after Heisenberg's, looked as different as could be. A late bloomer by the exacting standards of maths and physics at the time, Schrödinger had secured a professorship at the University of Zürich in 1925, aged thirty-seven. In October that year, he had latched on to the work of the French duke Louis de Broglie, who had proposed that particles such as electrons had both wave-like and particle-like properties.[12] The baffling 'wave–particle' duality that de Broglie espoused found few immediate converts – how could matter be both particle and wave? But experiments in 1927 would prove him right: a stream of electrons could be diffracted and made to interfere much like light, the principle behind the electron microscope.[13] Schrödinger, however, realized what was missing from de Broglie's work was an equation describing how matter waves snaked through space and time, similar to those derived for light (and other electromagnetic waves) by Scottish physicist James Clerk Maxwell in the nineteenth century.

Anxious to make his name with a big discovery, Schrödinger worked through a two-week tryst with an ex-lover in an Alpine resort that Christmas, returning to Zurich in January to apply his new wave equation to some of the key problems that were being thrown up by atomic physics. 'A late erotic outburst' was how Hermann Weyl, a close friend of Schrödinger (and his wife's lover), would describe the deluge of academic papers that were to follow. Among them was a complete description of the hydrogen atom spectrum based on his theory and a version of his equation that showed how the waves evolved over time.

What Schrödinger's mysterious waves really *were*, however, no one knew. Water waves and sound waves, for instance, are transmitted by the movement of water or air molecules. But what medium were 'matter waves' travelling through? Physicists rushed to embrace wave mechanics nevertheless, thankful that, unlike Heisenberg's matrix methods, Schrödinger's maths was reassuringly familiar and his equations often easier to solve.

In the case of the hydrogen atom, Schrödinger substituted into his equation values for the masses and charges of the electron and nucleus and a formula, from classical physics, for the electrical energy of the two particles. He then found functions that satisfied his equation – a process that could be accomplished by an undergraduate mathematician.[14] These 'wave functions', to which Schrödinger assigned the Greek letter psi (ψ), describe how the height of the wave (amplitude) varies in space and time. For the hydrogen atom, there are an infinite number of solutions to the Schrödinger equation, each representing one of Bohr's special orbits. The overall wave function for the atom, ψ, is an infinite sum or 'superposition' of them all.[15]

In the spring of 1925, there had been no theory that adequately described the physics of the atom. Less than twelve months later, there were two. Both theories seemed to do the job but were so different that many physicists wondered if they could both really be correct. Schrödinger confessed he was 'repelled'[16] by the instantaneous quantum jumps of Heisenberg's theory. When electron transitions took place in Schrödinger's theory, the wave function describing the atom changed smoothly from one form to another. Heisenberg was even blunter about the failings of Schrödinger's wave mechanics. 'It's crap,' he wrote to Pauli.[17] He was particularly disturbed by the physical picture that Schrödinger was trying to paint of the atom's inner workings. Heisenberg, who had avoided anything that could not be observed directly in his matrix mechanics, balked at the central role given to exotic, invisible 'waves' in Schrödinger's work. Born later showed that there was no easy physical interpretation of the wave function at all – it was a wave of probability, an ethereal entity that was not carried by or moving through anything.

Some physicists, satisfied there were now two formulations of quantum mechanics that appeared to give the right answers, shrugged off the unresolved questions swirling around quantum theory. Just choose the theory best suited to the problem at hand, they suggested, and hang the existential consequences. Even Heisenberg turned to Schrödinger's wave mechanics to calculate the spectrum of the helium atom. But Bohr, Einstein and other scientists were disturbed the two theories seemed to be saying very different things about the nature of reality. The more mathematically minded were equally uneasy that there was no straightforward way to reconcile the two theories. There had to be some deeper connection, they reasoned, between Heisenberg's infinitely large arrays of numbers and Schrödinger's strange formless waves of undulating probabilities. But what could it be?

At Göttingen, von Neumann heard about matrix mechanics first-hand. He was keen to help Hilbert extend his programme of axiomatization to physics. By sheer coincidence, both he and his hero were experts in the underlying mathematics of quantum theory.

In mathematical terms, Schrödinger was applying a mathematical 'operator' (the energy operator, known as the 'Hamiltonian') to his wave function to extract from it information about the energy of the system. Crudely speaking, operators are mathematical instructions. The solutions (i.e. the wave functions) to equations like Schrödinger's are 'eigenfunctions'. The answers (i.e. the energy levels of an atom) that pop out of the equation after eigenfunctions are substituted in are 'eigenvalues'. Hilbert himself had come up with these terms in 1904, based on the German word *eigen*, meaning 'characteristic' or 'inherent'. He had also pioneered spectral theory, which broadened the mathematics of operators and eigenvalues. A 'spectrum' in Hilbert's theory was the complete set of eigenvalues (i.e. solutions) associated with a particular operator. For example, the 'spectrum' of the Hamiltonian is, in the case of the hydrogen atom, the complete set of all allowed energy levels.

When Hilbert realized his two-decade-old spectral theory was proving useful in the dazzling new world of the quantum atom, he was delighted. 'I developed my theory', he said, 'from purely mathematical interests, and even called it "spectral analysis" without any presentiment that it would later find application to the actual spectrum of physics.' But the grand old man of mathematics, now in his sixties, was still confused by what he heard about quantum mechanics in Heisenberg's presentations.

Hilbert asked his assistant, Lothar Nordheim, to explain things to him but found the paper Nordheim produced to be unintelligible. When von Neumann saw it, he immediately realized that the deep mathematical structure of quantum theory could be recast into terms familiar to Hilbert. Nordheim's paper was his first clue to the essence of quantum theory, the common thread running through wave and matrix mechanics.

When wave and matrix mechanics appeared, many physicists suspected the missing link between the two theories would be discovered only by reconciling the two kinds of infinity that lay at their hearts. An atom has an infinite number of orbits, for example, so Heisenberg's matrices must also be of infinite size to represent all possible transitions between them. The members of such a matrix can, with sufficient patience, be lined up with a list of the counting numbers – they are 'countably' infinite.[18] Schrödinger's formulation, on the other hand, yielded wave functions describing in many instances an *un*countably infinite number of possibilities. An electron that is not bound to an atom, for example, could according to quantum theory be literally anywhere.[19] Until a measurement is made to determine its actual whereabouts, the electron is in a superposition of states, each corresponding to the electron being at some position (specified by coordinates x, y, z).[20] Heisenberg's matrices (with their countable elements) and Schrödinger's continuous wave functions are said to occupy different types of 'space'. 'Every attempt to relate the two,' von Neumann warns, 'must run into great difficulties' and 'cannot be achieved without some violence to the formalism and to mathematics'.[21]

One person who nonetheless tried to do exactly that was the taciturn British theoretical physicist Paul Dirac, whom novelist Ian McEwan describes as 'a man entirely claimed by science, bereft of small talk and other human skills'.[22] His Cambridge colleagues even named a unit of speech after him: a 'dirac' amounted to a single, solitary word per hour. Dirac would later fall in love with and marry Margit Wigner, the sister of Johnny's school friend Eugene. He would even learn to tell a joke or two. But in the 1920s, the young Dirac was a man who cared little for anything other than advanced physics and who, in the words of Freeman Dyson, 'seemed to be able to conjure laws of nature from pure thought'.[23]

Dirac began to expound his version of quantum theory in 1925.[24] In his 1930 book *The Principles of Quantum Mechanics*,[25] he set out an ingenious trick to merge the 'discrete' space of Heisenberg's matrices and the other 'continuous' space of Schrödinger's waves. The key to Dirac's approach was a special mathematical device that is now named after him: the Dirac delta function. This is a very peculiar entity indeed: everywhere except the origin, the function is equal to 0; but at the origin, it is infinitely high. The area under this vanishingly thin spike was defined by Dirac to be equal to one.

Dirac's function was forbidden by the rules of mathematics. He did not care. When Hilbert chided that the delta function could lead to mathematical contradictions, Dirac airily replied, 'Did I get into mathematical contradictions?'[26] Armed with his delta function, Dirac was able to show wave and matrix mechanics might after all be two sides of the same coin. The delta function acts as a sort of salami slicer, cutting up the wave function into manageable, ultra-thin slivers in space. If one accepts the use of Dirac's delta function, then the mathematical complications of reconciling wave and matrix mechanics appear to be magicked away. The wave function is chopped into bite-size chunks at every point in space. It seems that just as in matrix mechanics, there are then an infinite number of elements to contend with rather than a smoothly varying wave.

Like many mathematicians, von Neumann was dissatisfied with this imperfect union. He dismissed the delta function as 'improper', 'impossible' and a 'mathematical fiction'. He wanted a less sloppy take on the new science. The vital clue to von Neumann's rigorous reformulation of quantum mechanics lay in the early work of Hilbert.

Soon after Schrödinger unveiled his wave formulation of quantum physics, von Neumann, Dirac, Born and others realized that the mathematics of operators, eigenvalues and eigenfunctions could be useful in matrix mechanics too. Operators could be written as matrices.[27]

But operators have to act on something, and whereas in Schrödinger's theory they acted on wave functions, Heisenberg had made no reference to quantum 'states' in his early work because they could not be observed directly (he was working only with the intensities and frequencies of spectral lines). The concept was consequently introduced to matrix mechanics, with an infinitely long column or row matrix (i.e. a single vertical or horizontal lane of numbers) representing a state in Heisenberg's theory in much the same way that the wave function did in Schrödinger's.

A row or column matrix can be thought of as a vector, pointing to coordinates given by the numbers in the matrix. Since a state matrix is comprised of an infinitely long series of numbers, an infinite number of axes is required to represent this vector. This sort of infinite-dimensional space is impossible for anyone to picture. Nonetheless, the maths dealing with these daunting spaces had been set out by Hilbert in the first decade of the twentieth century, and von Neumann, who now quickly made himself the world's leading expert on the subject, named them 'Hilbert spaces' in honour of his mentor.[28]

By definition, to be a proper Hilbert space, the squares of each number comprising a vector added together has to be finite.[29] Hilbert was exploring these spaces because they are mathematically interesting, and all sorts of results from school geometry (such as Pythagoras' theorem) apply. Crucially, Hilbert spaces can be also formed by certain sets of functions as well as numbers. One class of function that had been shown to form a Hilbert space were those that are square integrable – squared and summed over all space, such functions are finite.

The quantum wave function is just such a function. Born had shown that the square of the amplitude of the wave function at any point indicated the chance that a particle will be found at that particular position. Since it is certain that a particle must be *somewhere* in space, the wave function squared and summed over all space must be 1. That means quantum wave functions are square-integrable and form a Hilbert space.[30]

Von Neumann may have lacked Dirac's intuitive, almost mystical, physical insight but he was a far better mathematician. In 1907, mathematicians Frigyes Riesz and Ernst Fischer had, within months of each other, independently published proofs of an important result relating to square integrable functions, and von Neumann realized that their work could connect wave and matrix mechanics. Square integrable functions such as the wave function can be represented by an infinite series of orthogonal functions,[31] sets of mathematically independent functions that can be added together to make any other.[32] Imagine having to fill a 124-litre trough exactly to the brim with 20-, 10- and 7-litre buckets. One way to do it would be with five bucketfuls of the first, one of the second and two of the third. A wave function can be similarly 'topped up' by adding up bits of other functions. How much of each function is required is indicated by their coefficients.[33] What Riesz and Fischer showed was that if the square of the wave function is 1, then the sum of each of these coefficients squared is 1 too.[34]

Armed with this theorem, von Neumann quickly spotted the link between the Heisenberg and Schrödinger theories: the coefficients of the expanded wave function were exactly the elements that appear in the state matrix. According to Riesz-Fischer, the two seemingly disparate spaces are in fact the same; and the two spaces were, as von Neumann put it, 'the real analytical substrata of the wave and matrix theories'.[35] Giants of quantum theory like Dirac and Schrödinger had tried to prove the equivalence of the two. Von Neumann was the first to crack it, showing decisively that wave and matrix mechanics were fundamentally the same theory. But never before had two descriptions of the same phenomena implied such different pictures of reality. Newton's gravitational law described how planets wheeled through the heavens, the kinetic theory of gases assumed the motion of huge numbers of particles accounted for their properties, but what, if anything, did the mathematics of the quantum theory represent? Von Neumann had built a rock in the midst of a sea of possibilities.

Von Neumann's stay in Göttingen was brief, though he was to visit many more times over the next few years. When his Rockefeller fellowship ran out in 1927, he was offered a job at the University of Berlin. He was the youngest *Privatdocent* the university had ever appointed. The position gave him no salary, only the right to lecture and receive fees directly from students. But there were distractions. The German Empire had collapsed after the end of the First World

War. Berlin was now the wild, decadent capital of the Weimar Republic. A popular ditty among Berliners at the time, ran:

> Du bist verrückt, mein Kind,
> Du mußt nach Berlin,
> Wo die Verrückten sind,
> Da jehörst de hin.
>
> You are crazy, my child,
> You must go to Berlin,
> That's where the crazy are,
> That's where you belong.[36]

The twenty-three-year old lapped it up. The bookish Wigner was also in Berlin but apart from socializing with his fellow Hungarians (Teller, Szilard and von Neumann) and attending the lively physics colloquia, he led a rather monastic existence. Von Neumann, Wigner recollected, lived a very different sort of a life. 'He was sort of a bon vivant, and went to cabarets and all that.'[37]

Friedrich-Wilhelms-Universität Berlin

Ausweiskarte

Inhaber dieser Karte
Privatdozent
Dr. Johann Neumann
von Margitta
ist Angehöriger der Friedrich Wilhelms-Universität zu Berlin und hat Zutritt zu dem Universitäts-Gebäude.

Eigenhändige Unterschrift des Inhabers

John von Neumann's identity card, issued by the University of Berlin. *Courtesy of Marina von Neumann Whitman.*

As well as the vibrant nightlife of Berlin, the scientific culture was second to none. German, not English, was the language of science in the 1920s. Practically all the founding papers of quantum mechanics were written in it. There was a flood of congresses and conferences for young researchers to attend. Academic talks would often spill over into coffee houses and bars. 'The United States in those years was a bit like Russia: a large country without first-rate scientific training or research,' Wigner told an interviewer in 1988. 'Germany was then the greatest scientific nation on earth.'

Von Neumann's usual approach to giving seminars was not to spoil them by over-preparing. He would often think through what he might say on the train journey to the conference, turn up at the seminar with no notes and then race through the maths. If he filled up the blackboard, he would rub out a swathe of earlier equations and plough on. Those not as quick on the uptake as he was – i.e. nearly everyone – referred to his inimitable seminar style as 'proof by erasure'. Any tensions that arose, however, he could, and often did, defuse by telling risqué jokes in any of three different languages. When someone else's presentation bored him, he would look engrossed while mentally retreating from the room to think about other, more interesting mathematical problems.

Von Neumann enjoyed his Berlin years immensely but he realized the chances of securing a paid professorship would be better elsewhere. He took up the offer of a job at Hamburg in 1929, hoping to be promoted quickly to a full professorship. He would not be there long.

Meanwhile he was busy reducing the whole of quantum mechanics to its mathematical essentials, just as he had set theory. First working with Nordheim and (nominally) Hilbert, then later on his own, von Neumann developed the thinking that would culminate in his 1932 masterpiece of mathematical physics, *Mathematical Foundations of Quantum Mechanics*, which showed how quantum theory emerged naturally by considering the mathematical properties of Hilbert space.[38] Satisfied that he had produced the most rigorous formulation of quantum theory, he turned his attention to the most contentious question of the day in physics: what on earth was going on beneath all that elegant maths?

Physicists have wrestled with what quantum mechanics is really telling us about the nature of the physical world since its early days. The failure to come up with an acceptable interpretation of the theory even led students at Schrödinger's university to make up a ditty gently ribbing their great professor:

> Erwin with his psi can do
> Calculations quite a few,
> But one thing has not been seen:
> Just what does psi really mean?[39]

The existence of GPS, computer chips, lasers and electron microscopes attest that quantum theory works beautifully. But nearly a hundred years after Heisenberg published his paper on matrix mechanics there is still no agreement on its meaning. In the interim, a plethora of exotic ideas have been put forward to make sense of what quantum physics is saying about reality. All have their passionate advocates, but none have yet been proven. Physicists ruefully joke that though new interpretations of quantum physics arrive with astonishing regularity, none ever go away. For many, that joke is turning sour. 'It is a bad sign', theoretical physicist Steven Weinberg noted recently, 'that those physicists today who are most comfortable with quantum mechanics do not agree with one another about what it all means.'[40]

At the heart of the problem is the borderland between quantum and classical physics, where the interactions of atoms and photons are revealed to us via instruments like microscopes and spectrometers or our own eyes. According to quantum theory, a particle can be in a superposition of infinitely many states. In the case of a free electron, for example, the particle is anywhere and everywhere and its wave function is a superposition of states representing all these possibilities.

Now imagine that we 'catch' an electron on a phosphor screen. When an electron collides with the screen, the phosphor coating emits a shower of photons, and we see a flash, telling us the approximate position of the electron. The electron is now in one state with a single corresponding (eigen)value for its position. At no point does an observer 'see' the electron's wave function splayed improperly across all space. The electron is either in a superposition of states that are inaccessible to an observer – or, after observation, localized at a point. As if embarrassed by its own naked quantumness, the particle appears to have donned classical clothes the moment it is observed.

The two situations, before and after an observation, are completely different, and this duality, von Neumann says, is 'fundamental to the theory'. The particle is at first described by a wave function (comprised of all possible states). This wave function is a solution of the Schrödinger equation, perfectly describing the particle anywhere in space and at any point in time. Just as a satellite's orbit around the Earth can be calculated from moment to moment thanks to Newton and Einstein's equations, Schrödinger's equation allows the evolution of the wave function to be known precisely anywhere in space and time. This behaviour is as deterministic as Newton's laws of motion. But when we try to find out something about a particle, like its position or momentum, the wave function pops like a bubble and the particle adopts, at random, a state out of all the available possibilities. This process is discontinuous and cannot be reversed: once the particle has 'selected' a particular state, Schrödinger's equation no longer holds, and the other states are lost as possibilities. Now known as 'wave function collapse', this process is unlike anything in classical physics.

In *Mathematical Foundations of Quantum Mechanics*, von Neumann credits Bohr with first identifying these two essentially incompatible processes in 1929. But Bohr's rambling essay of that year rather obscures the issue.[41] He suggests that the measuring instrument (the microscope, for example), a large 'classical' object, intruding into the quantum world, is somehow necessary for the irreversible change that occurs during an observation. As von Neumann notes, however, there is no clear-cut divide between the classical and quantum realms. Measuring devices are, after all, made of atoms that obey the laws of quantum physics. There is nothing in the maths to say if or when a collection of atoms gets 'big' enough to pop the wave function.

The question of how, when or even if, the wave function collapses is at the root of the so-called 'measurement problem', and the differences between the plethora of interpretations of quantum mechanics today usually hinge on their respective answers to this question, first dissected and analysed thoroughly in von Neumann's book of 1932.

Von Neumann's discussion of measurement begins simply: he imagines measuring the temperature of something (a cup of coffee, say) using a mercury thermometer. This requires at least one person, an observer, to see where the column of mercury has risen to on the thermometer's scale. Between thermometer and observer, von Neumann argues, one can insert any number of processes.

The light entering the observer's eye, for example, is a stream of photons that have been reflected by the mercury column, and refracted by the observer's eye before hitting the retina. Next, the photons are transformed by retinal cells into electrical signals that travel up the optic nerve to the brain. Furthermore, the signals from the optic nerve might be expected to induce chemical reactions in the brain. But no matter how many such steps we add, von Neumann argues, the sequence of events must end with *someone* perceiving these events. 'That is,' he says, 'we are obliged always to divide the world into two parts, the one being the observed system, the other the observer.'

But what about the steps in between? The most straightforward interpretation of quantum mechanics would seem to require that any number of such steps would lead to the same result – at least as far as the observer is concerned. If this were not so, then the way that someone chose to slice up the problem would give different predictions about what the observer sees. No one wants a theory that gives different answers for what is, in essence, the same problem. So von Neumann worked out if quantum mechanics did indeed give the same answer for all scenarios that start and end in the same way.

To do so, von Neumann divided the world into three parts. In the first case, part I is the system being measured (the cup of coffee); part II, the measuring device (thermometer); and part III, the light from the device and the observer. In the second case, part I is the cup and the thermometer; part II includes the path of light from the thermometer to the observer's retina; and part III, the observer from the retina onwards. For his third example, von Neumann considered the situation where part I encompasses everything up to the observer's eye; II, the retina, optic nerve and brain; and III, what he called the observer's 'abstract "ego"'. He then calculated for his three examples the consequences of putting the boundary (the point at which the wave function collapses) between part I and the rest of the experiment. Next he shifted the boundary so that wave function collapse occurred after parts I and II but before III and recalculated the outcomes from the observer's point of view.

To do the maths, von Neumann needed to work out what happened in quantum theory when a pair of objects interact. In this situation, he found the quantum states of the coffee cup and thermometer, for instance, can no longer be described independently of each other or even as a superposition of their individual states. According to his formalism, their wave functions become so inextricably intertwined that both must be represented together by a single wave function. Schrödinger would in 1935 coin the term 'quantum entanglement' to describe this phenomenon. This means that measuring some property of one of the pair instantly collapses the wave function of the whole system, even if the objects are separated by some vast distance after their initial interaction. Einstein, who was probably the first to fully appreciate this consequence of entanglement and did not like it one bit, called it 'spooky action at a distance'.[42]

Von Neumann was always rather more relaxed about the weirder aspects of quantum physics than Einstein. He wanted to know if the duality inherent in the new quantum physics meant the theory would contradict itself. Reassuringly, he found it did not. Wherever he put the dividing line between quantum and classical, the answer, as far as an observer was concerned, was the same. 'This boundary,' he concluded, 'can be pushed arbitrarily far into the interior of the body of the actual observer.' And that was true, said von Neumann, right up until the act of perception (whatever *that* was). The 'boundary' that he describes is now known as the 'Heisenberg cut'. More rarely (but perhaps more fairly) it is called the Heisenberg-von Neumann cut.

Von Neumann's results implied that in principle anything could be treated as a quantum object, whatever its size or complexity – as long as wave function collapse occurred (instantly) at some point in the chain between the system being observed and the consciousness of the person doing the observing. In this picture, it makes no sense to talk about the properties of an object (whether it be a photon, coffee cup or thermometer) until a measurement is made. None of these objects could be said to *be* somewhere in particular, for example, unless their wave functions had collapsed. This would become a fundamental tenet of the 'Copenhagen interpretation', for many years the prevailing view of what quantum mechanics means.[43] According to Copenhagen, the theory does not tell us what quantum reality 'is', only what can be known. Many physicists were attracted to it as it allowed them to get on with the business of physics without getting bogged down in speculations about things they could not see (no coincidence that Heisenberg, who rejected unobservable phenomena in his original formulation of matrix mechanics, promoted this view). Others felt the approach side-stepped the bigger questions posed by the theory. 'Shut up and calculate' was how the physicist David Mermin summarized the Copenhagenists' approach in 1989.[44]

Some of the most eminent founders of quantum mechanics were not altogether happy with the emerging consensus that von Neumann was helping to build. What, for instance, caused the wave function to collapse? Von Neumann did not tackle this problem head-on in his book. Others, including his friend Wigner, who often discussed such things with him, would later suggest that the consciousness of the (human) observer was responsible – a conclusion implied but not stated overtly in von Neumann's work.[45] Einstein strongly objected to this idea – the Dutch physicist and historian Abraham Pais recalled 'that during one walk Einstein suddenly stopped, turned to me and asked whether I really believed that the moon exists only when I look at it'.[46] Einstein (and he was hardly alone) felt that things should have properties regardless of whether there was someone there to see them.

Mathematical Foundations of Quantum Mechanics is the work of an exceptional mathematician. One of the work's earliest fans would be a teenager, who ordered the book in the original German as his prize after winning a school competition.[47] 'Very interesting, and not at all difficult reading' was how Alan Turing described von Neumann's classic in a letter to his mother the following year.[48] But von Neumann's book was also the work of a rather cocksure young man. To some, it seemed like the twenty-eight-year-old upstart was suggesting his book was the last word on quantum mechanics.

Erwin Schrödinger disagreed. Three years after the publication of von Neumann's book, Schrödinger discussed with Einstein the weaknesses in what would become known as the Copenhagen interpretation of quantum mechanics. Inspired by their frenzied exchange of letters, Schrödinger posed the most famous thought experiment of all time to highlight the absurdity of applying quantum mechanics willy-nilly to everyday objects.[49] If the rules of quantum mechanics could, as von Neumann argued, be applied just as well to large things, then why, thought Schrödinger, should they not apply to insects? Or mice? Or cats?

'One can even set up quite ridiculous cases,' wrote Schrödinger in his 1935 paper.

> A cat is penned up in a steel chamber, along with the following diabolical device (which must be secured against direct interference by the cat): in a Geiger counter there is a tiny bit of radioactive substance, so small, that perhaps in the course of one hour one of the atoms decays, but also, with equal probability, perhaps none; if it happens, the counter tube discharges and through a relay releases a hammer which shatters a small flask of hydrocyanic acid. If one has left this entire system to itself for an hour, one would say that the cat still lives *if* meanwhile no atom has decayed. The first atomic decay would have poisoned it. The psi-function of the entire system would express this by having in it the living and the dead cat (pardon the expression) mixed or smeared out in equal parts.

Schrödinger's cat was a gotcha of the highest order, a takedown of efforts to paper over the cracks in quantum theory. A cat, most people would agree, can be either dead or alive. But if we follow von Neumann's logic, until someone opens the chamber, the cat's wave function is entangled (the term is used for the first time in this paper) with that of the radioactive substance, and the unfortunate feline is both alive *and* dead. If quantum mechanics can result in such patently obvious nonsense at the macroscopic scale, how can we know the theory 'truly' describes the atomic realm? Schrödinger was intimating that quantum theory was not the end of the road. 'The theory yields much,' Einstein famously wrote to Born, 'but it hardly brings us closer to the Old One's secrets. I, in any case, am convinced that He does not play dice.'[50] Like Einstein, Schrödinger felt there must be another, deeper theory underlying quantum mechanics that would provide a more sensible physical picture of what was going on. The moon exists, even if there is no one to see it. An electron must have properties – be *somewhere*, for example – before it is caught on a phosphor-coated screen. In what has become the most controversial part of his book in recent years, von Neumann discusses this idea – and seemingly dismisses it.

Quantum mechanics, as we have seen, is strikingly different from the physical theories that preceded it. If the Copenhagen interpretation is correct, then the collapse of the wave function results in an unpredictable outcome. A particle that is observed adopts a state at random out of all those available. This means that quantum theory is neither causal (we can't trace back exactly the events that result in a particle ending up where we observe it to be) nor deterministic (because the outcome of a particular observation is determined in part by chance). One way to restore causality and determinism to the quantum world, and realism to boot (so particles have properties even if no one measures them), is to assert the existence of 'hidden variables' or 'hidden parameters': properties that are associated with all particles but inaccessible to observers.[51] In such schemes, these unobservable parameters completely determine the state of the system. The element of chance is eliminated: no dice-playing God is required.[52] Von Neumann was sceptical that a theory based on hidden variables could reproduce all the predictions of quantum physics. In the *Mathematical Foundations* he sets about demonstrating the very great difficulties that a hidden variable theory would encounter in doing so.

Imagine making some sort of measurement on an ensemble of many quantum particles (hydrogen atoms, say). Now make the same measurement on another, identical ensemble. In keeping with quantum theory and countless experiments, the measurements give different results. If the same measurement is made on a huge number of ensembles, then you find the results are distributed across a range of values. Any collection of particles displaying this sort of statistical variation is called a *dispersive* ensemble and so, according to quantum physics *all* ensembles are dispersive.

There are two possible reasons why an ensemble might be

dispersive, von Neumann says. One explanation might be that while the ensembles *seem* identical, the values of the hidden variables associated with the particles of each ensemble are different and, in sum, these unobservable parameters (which differ from ensemble to ensemble) account for the range of results obtained from measurements. This means that the ensembles cannot be composed of particles that *all* have the same hidden variable values (otherwise a measurement on an ensemble would always give the same result); in physics terms, they cannot be *homogeneous*. A second explanation is that contemporary quantum theory is correct, and the results of measurement are randomly distributed (so hidden variables are not necessary).

What von Neumann then proceeds to prove is that dispersive ensembles in quantum mechanics *are* homogeneous. All of the particles in the ensemble are in the same quantum superposition of states until there's a measurement. As he has already shown that hidden variables would mean that ensembles cannot, in general, be homogeneous, von Neumann can rule them out.

Von Neumann's proof electrified those who were leaning towards the Copenhagen interpretation. As word spread that the young genius had decisively rejected hidden variable theories, 'Von Neumann was hailed by his followers and credited even by his opponents,' says the historian Max Jammer.[53] By this time, von Neumann was enjoying the comforts of a new life in the United States.

Towards the end of October 1929, Wigner had received, out of the blue, an offer of a one-term lectureship from Princeton University. If that were not enough, the telegram quoted a salary so high – more than seven times what Wigner was earning in Berlin – he thought there must have been a mistake during the message's transmission. He quickly learned that von Neumann had received a letter from Princeton a couple of weeks earlier with the offer of even more money. 'It was clearly Jancsi that Princeton really wanted,' said Wigner.[54] Unbeknownst to him, von Neumann's letter asked if Wigner should be invited too. Luckily for Wigner, von Neumann agreed this would be a good idea. He added that there would, however, be a short delay before he could take up the post because he wanted to 'fix a family matter'. Von Neumann was going to Budapest to get married.

The scheme to entice the two Hungarians had been cooked up by Oswald Veblen, a distinguished Princeton professor of mathematics. America was an intellectual backwater and Veblen wanted to change that by poaching some of Europe's most brilliant mathematicians with the offer of huge American salaries. He had secured millions of dollars from the Rockefeller foundation and wealthy private donors to erect a grand new building, named Fine Hall, for the mathematics department. Now he just needed the mathematicians to fill it. Veblen came under pressure from the wider faculty to hire a physicist. Von Neumann and Wigner had recently co-authored papers on the puzzling spectra of atoms more complex than hydrogen. So Veblen came up with the perfect compromise: invite both Hungarians for half a year.

Wigner's boat arrived in New York harbour in early January 1930. Von Neumann arrived with his new wife, Mariette Kövesi, about a day later.[55] 'I met him, and we spoke Hungarian,' said Wigner. 'We agreed that we should try to become somewhat American.' On that day, Jancsi became Johnny, and Jenő became Eugene. 'Jancsi felt at home in America from the first day,' Wigner continued. 'He was a cheerful man, an optimist who loved money and believed firmly in human progress. Such men were far more common in the United States than in the Jewish circles of central Europe.'

Within a year or two of the publication of his book, von Neumann's 'impossibility proof' became gospel in the world of quantum theory. For decades, any young physicist keen to advance their careers would think twice before venturing to work on an alternative to the prevailing theory. 'Many generations of graduate students who might have been tempted to try to construct hidden-variables theories,' said physicist David Mermin in 1993, 'were beaten into submission by the claim that von Neumann, 1932, had proved that it could not be done.'[56]

But what, exactly, had von Neumann proved? The problem was that by dint of his reputation – and the fact that *Foundations* was not translated into English for a further two decades – few would closely scrutinize the proof itself. One person who did, two years after von Neumann's book was published, was the German mathematician and philosopher Grete Hermann.

Hermann had studied mathematics at Göttingen, and that she got there at all was already something of an achievement: girls were not generally admitted to the gymnasium she attended, and she had needed special dispensation to start her schooling. After graduating from the university, she became the only female doctoral student of the only female professor of mathematics there, the brilliant Emmy Noether. Just a few years earlier, historians and linguists at the university had tried to block Noether's own appointment, forcing Hilbert to intervene on her behalf. 'I do not see that the sex of the candidate is an argument against her admission,' he retorted. 'We are a university, not a bath house.' The hostility helped forge a bond between the two women, and Hermann would remember Noether fondly in her memoirs. When, after passing her doctoral examination in February 1925, Hermann announced her intention to pursue philosophy, Noether, who was in the midst of finding a job for her at the University of Freiburg, was not pleased: 'She studies mathematics for four years, and suddenly discovers her philosophical heart!'[57]

Hermann was as passionately committed to socialism as she was to the philosophy of Immanuel Kant. She joined the International Socialist Militant League, a part of the German resistance movement. She eventually fled to London and by becoming a British citizen through a marriage of convenience avoided being interned by the authorities. She returned to Germany after the war to help with reconstruction and was a harsh critic of intellectuals who had chosen to live and work under the Third Reich.

Sometime in 1934, she travelled to the University of Leipzig, where Heisenberg was a professor, to defend Kant's conception of causality from the onslaught of quantum theory. 'Grete Hermann believed she could prove the causal law – in the form Kant had given it – was unshakable,' Heisenberg wrote later. 'Now the new quantum mechanics seemed to be challenging the Kantian conception, and she had accordingly decided to fight the matter out with us.'[58] Heisenberg was impressed enough to devote a full chapter of his autobiography to Hermann's arguments.

Shortly after her time in Leipzig, Hermann published her critique of von Neumann's impossibility proof as part of a longer paper on quantum mechanics. She had identified a weakness in one of his assumptions, 'the additivity postulate', which she argued meant the proof was circular.[59] Essentially, Hermann said, von Neumann had shown his Hilbert space perfectly explained quantum physics and had then assumed any theory must have the same mathematical structure. But, she continued, if in future a hidden variable theory was discovered that *could* account perfectly for everything in quantum mechanics, there was no reason at all to assume it would resemble von Neumann's.

Hermann had in 1933 sent an earlier essay containing her discussion of the impossibility proof to Dirac, Heisenberg and others to prepare the ground.[60] That did not prevent her 1935 paper sinking into obscurity. [61] Hermann herself did not appear to attach much importance to it: she did not include the argument against von Neumann in the abridged version of the paper that was published by the prestigious journal *Naturwissenschaften*.[62] Quite possibly she felt that rigorous philosophy, not more mathematics, was required to save determinism.[63]

Not until 1966, thirty years after Hermann had published her critique, were the limitations of the impossibility proof to become more widely known. 'The von Neumann proof, when you actually come to grips with it, falls apart in your hands!' John Stewart Bell was to declare many years later. 'There is *nothing* to it. It's not just flawed, it's *silly*!'[64] Born in Belfast into a poor family, Bell was the only one of four siblings to stay at school after the age of fourteen. He worked as a technician in the physics department at Queen's University, where after a year he secured a small scholarship and went on to get a degree in experimental physics and one in mathematical physics a year later in 1949. Feeling guilty for being dependent on his parents for so long, Bell immediately found a job at the Atomic Energy Research Establishment at Harwell. He would only be able to pursue a PhD years later, getting his doctorate in 1956. Four years later he moved to CERN in Geneva with his wife, fellow physicist Mary Bell, where he worked on particle physics and the design of particle accelerators. 'I am a quantum engineer, but on Sundays I have principles,' he proudly told a gathering of PhD students during an impromptu seminar in 1983.[65] The principles that Bell was interested in were, of course, those of quantum mechanics.

Ever since Bell had first studied physics, he had felt that something was 'rotten' in quantum physics. What bothered him most was the duality von Neumann had identified, the movable 'cut' between the quantum and classical worlds. He was attracted to the idea of hidden variable theories because they could in principle make this boundary disappear. The quantum lottery of wave function collapse would be unnecessary, and there would be a smooth transition from quantum to classical or, in Bell's words, 'a homogeneous account of the world'. Still Bell accepted what was by then the consensus: von Neumann's proof precluded hidden variable theories. Bell could not read German and an English translation of von Neumann's book would not be available for another three years. He accepted the second-hand accounts he read.

That changed in 1952. 'I saw the impossible done,' Bell said.[66] In two papers the American physicist David Bohm had described a hidden variable theory that could reproduce the results of quantum mechanics in their entirety. The Copenhagen interpretation was by now orthodoxy. Bohm's scheme was heresy, but he was an outsider with little to lose. He had been hauled up in front of the House Committee on Un-American Activities in 1949 for his communist affiliations, then arrested for refusing to answer questions. Though he was acquitted, Princeton refused to reinstate him, and he was advised to leave the United States by his former doctoral adviser, Robert Oppenheimer. Bohm took the hint, accepting the offer of a professorship at the University of São Paulo. His theory was the work of exile.

Bohm had cleverly modified Schrödinger's equation so that the wave function is transformed into a 'pilot wave'. The trajectories of particles are guided by this wave such that their behaviour was in keeping with the rules of quantum mechanics. Any physical changes that might affect a particle, no matter how far away they occurred, are instantly transmitted by the pilot wave, which pervades the whole universe. If all the factors affecting a particle are known, then in principle its path can be calculated exactly from beginning to end. Bohm's theory allows determinists to have their quantum cake and eat it.

Bell could not believe it. 'Why is the pilot wave picture ignored in text books?' he asked in 1983. 'Should it not be taught, not as the only way, but as an antidote to the prevailing complacency?'[67] But Bell was busy with his day job. Only in 1964, during a year-long sabbatical at the Stanford Linear Accelerator Center in California, did he return to the impossibility proof, now with an English translation of von Neumann's book. At Stanford, he independently discovered the same flaw that Hermann had highlighted so many years earlier. The resulting paper eventually appeared in 1966.[68] (The cause of the delay was that a letter sent to Stanford from the journal's editor was not forwarded to Bell, who had returned to CERN.) But Bell's work was published by *Reviews of Modern Physics*, one of the most prestigious journals in the field. He was also, quite soon, to become famous. His refutation of von Neumann's work did not suffer the same fate as that of Hermann. An interest in the meaning of quantum mechanics was soon no longer necessarily career suicide. Liberated, some physicists began to examine the foundations of quantum theory, as they had in the 1920s, when it was born. The end of the Copenhagen hegemony was nigh, and new interpretations began to spring up like weeds.

Debate still rages among quantum physicists over von Neumann's 'impossibility proof'. Mermin is among those who believe that von Neumann erred and Bell and Hermann correctly identified his error.[69] Jeffrey Bub and, separately, Dennis Dieks have argued that von Neumann never meant to rule out all possible hidden variables – only a subset of them.[70] In essence, they say that all von Neumann was aiming to prove is that no hidden variable theory can have the same mathematical structure as his own; they cannot be Hilbert space theories. And that is certainly the case: Bohm's theory, for example, is quite different from von Neumann's.

While Heisenberg and Pauli branded Bohm's theory as 'metaphysical' or 'ideological', von Neumann was not dismissive, as Bohm himself notes with some pride and more than a little relief. 'It appears that von Neumann has agreed that my interpretation is logically consistent and leads to all results of the usual interpretation. (This I am told by some people.)' Bohm wrote to Pauli shortly before his theory was published. 'Also, he came to a talk of mine and did not raise any objections.'[71]

Bohm might have hoped for Einstein to embrace his ideas, which restored both realism (particles exist at all times in Bohmian mechanics) and determinism. Einstein was, however, less kind than von Neumann. Disappointed that Bohm had not rid quantum mechanics of 'spooky action at a distance' (which he could not abide) he privately called Bohm's theory 'too cheap'.[72] He had been similarly unimpressed a quarter of a century earlier, when de Broglie had presented a nascent version of the pilot wave picture. Einstein, who plumbed the quantum's depths, never devised a satisfactory alternative of his own.

Even Bell, who championed Bohm's approach, had reservations. 'Terrible things happened in the Bohm theory,' he conceded. 'For example, the [paths of] particles were instantaneously changed when anyone moved a magnet anywhere in the universe.' Bell wanted to explore this aspect of Bohm's theory further. The same year that he wrote his paper criticizing von Neumann's impossibility proof, he was also working on another, exploring whether any theory could account for the results of quantum mechanics without appearing to need, like Bohm's, some sort of instantaneous signalling between particles separated by vast distances. This 'nonlocality' is built into both standard quantum mechanics (via entanglement) and Bohm's theory (via the all-seeing pilot wave). And that looks awkward at first glance because according to Einstein's special theory of relativity (which no experiment has ever contradicted) nothing can travel faster than light. Einstein, with two collaborators, Nathan Rosen and Boris Podolsky, had pointed this out himself with a famous thought experiment that became known as the EPR paradox.[73] The upshot of their paper was that quantum theory must be incomplete because, according to the theory, a measurement of a particular property on one of a pair of entangled particles immediately determines the corresponding state of the other, no matter how far away it is. Since no signal between the two can travel faster than light, the authors reasoned, these values must somehow be fixed before measurement and not determined by the act of measurement, as quantum theory dictates.

In fact, as is now widely appreciated, quantum theory does not violate special relativity. A measurement on one of a pair of entangled particles does not directly affect the state of the other; there is only a correlation between the two and no causal link. No message can ever be sent faster than light by means of an entangled pair because to understand the message, its recipient would have to know the result of the measurement on the sender's particle. There is no 'spooky action at a distance', as Einstein feared, because there is no 'action'. But what Bell wondered was: could any 'local' hidden variables theory account for the correlations between the two particles that quantum theory attributed to entanglement? What if, as Bell would put it later, there was no weird quantum entanglement but just a case of Bertlmann's socks? Reinhold Bertlmann, Bell's friend and collaborator, always wore socks of different colours. 'When you see that the first sock is pink you can be already sure that the second sock will not be pink,' wrote Bell. 'And is not the EPR business just the same?'

If that were the case and hidden variables had determined the relevant properties prior to any measurements, there would be no need for the strange ideas of the Copenhagenists. Was it possible to differentiate between the two possibilities? Bell's genius was to realize that it was.

He pictured a simpler, more practical version of the EPR thought experiment that had been devised by Bohm. In this, two entangled particles are created and fly apart until the distance between them means they cannot communicate at slower-than-light speeds in the time it takes to perform a measurement on them. Bohm's suggestion was to measure spin, a quantum property of subatomic particles like electrons and photons. A particle can either be 'spin-up' or 'spin-down'. He proposed splitting a hydrogen molecule, with no spin, into a pair of hydrogen atoms. Since the total spin of the two atoms must still be zero, one must be spin-up and the other spin-down. If the orientations of the two spin detectors are aligned, then this is exactly the result one gets, 100 per cent of the time.

Bell's idea was to vary the relative orientations of the two detectors so that there was an angle between the two measurements. Now if one of the two particles is measured as having spin-up, the other one is not always spin-down. According to quantum theory, however, the fates of the two particles are still linked and, as a result, the outcome of one spin-up and the other spin-down is still strongly preferred. What Bell showed mathematically is that, for certain orientations of the two detectors, the correlation between the spins of the two particles would have to be lower, on average, for a local hidden variable theory than quantum theory. Bell's theorem takes the form of an inequality that places a limit on how high such correlations can be for any local hidden variables theory. Any correlation higher than this limit is said to 'violate' Bell's inequality and would mean either quantum theory or some other non-local theory like Bohm's must be at work. The laser technology required to probe a particle's spin soon improved enough to allow Bell's theorem to be experimentally tested.[74] Physicists John Clauser and Stuart Freedman at the University of California at Berkeley carried out the first Bell test in 1972. Their experiment and the dozens since have all found a violation of Bell's inequality – a result that only quantum theory or a non-local theory like Bohm's can explain.

While Bohm's theory struggled to win widespread acceptance (though Bell continued to champion it), another was to be ignored altogether, only to spawn countless science fiction stories and half-baked mystical philosophies (and more than a few research papers) when it re-emerged over a decade later.

The progenitor of the 'Many Worlds' interpretation was a young American theorist named Hugh Everett III, who began his graduate studies at Princeton University in the mathematics department. By coincidence, he spent his first year working on game theory, a field which von Neumann had helped found with his 1944 book, *Theory of Games and Economic Behavior*. Soon, however, he took courses in quantum mechanics and, in 1954, mathematical physics with von Neumann's old friend Wigner. Von Neumann's book on quantum theory didn't come out in English until the following year, but his ideas were already well known in the United States, according to Everett. Von Neumann's formulation is 'the more common (at least in this country) form of quantum theory,' he says in a letter.[75] But Everett did not swallow the gospel whole.

Like Bell, who was also working towards his PhD on the other side of the Atlantic, he was dissatisfied with von Neumann's approach to the measurement problem. The abrupt transition from quantum to classical that is implied by wave function collapse is 'a "magic" process', he wrote to Jammer in 1973, quite unlike other physical processes, which 'obey perfectly natural continuous laws'. The 'artificial dichotomy' created by the 'cut', he said, is 'a philosophic monstrosity'.

Everett first hit upon his solution 'after a slosh or two of sherry' with his flatmate Charles Misner and Aage Petersen, Bohr's assistant, who was visiting Princeton at the time.[76] Von Neumann's approach had been to distil from the physics the bare mathematical principles required to explain quantum phenomena, then, using only these laws, infer whatever one could about the nature of the quantum realm. But

Everett realized von Neumann had not done that in his treatment of measurement. Instead, von Neumann had noted that observers never see a quantum superposition of states, only a single classical state. Then, he had *assumed* that at some point, a transition from quantum to classical must take place. Nothing in the maths necessitated wave function collapse. What, Everett wondered, if we *really* follow the maths to its logical conclusion? What if there is no collapse at all?

Everett was led to a startling result. With no artificial boundary to constrain it, there is quantumness everywhere. All the particles in the universe are entwined in a single massive superposition of all possible states. Everett called this the 'universal wave function'. Why, then, does an observer perceive only one outcome from a measurement and not some quantum fuzz of possibilities? That is where the 'many worlds' come in. Everett proposed that every time a measurement is made, the universe 'splits' to create a crop of alternative realities, in which each of the possibilities play out. (So Schrödinger's cat is alive in one universe and dead in another. Or in several.) One of many objections raised to Everett's ideas is that universes multiply like rabbits, a consequence that strikes some physicists as using an ontological sledgehammer to crack an epistemic nut.[77] In some versions of the theory 'measurement' can mean *any* quantum interaction. Every time a nucleus emits an alpha particle or a photon interrogates an atom, a whole new universe springs into being.

Impressed with the mathematical lucidity of Everett's thesis (if not entirely convinced by its substance), his PhD adviser, John Wheeler, took it to Copenhagen in the hope of winning Bohr's approval. He failed. Everett, disappointed with the reception his theory received from the physics community, left academia to work in weapons research for the Pentagon. But the Copenhagenists' grip on quantum physics was slowly loosening. In 1970, the American physicist Bryce DeWitt wrote an article for *Physics Today*, the membership magazine of the American Institute of Physics, and the theory began to make its way into the popular imagination, boosted by an article in the science fiction magazine *Analog*.

A welter of interpretations bloomed after Everett and Bohm highlighted some of the inadequacies of Copenhagen in the 1950s. The questions they raise are no longer of purely academic interest. Quantum mechanics now underlies a host of modern technologies, from fibre optics to microchips. The latest development is the quantum computer – still in its infancy, but potentially able to harness the power of quantum superpositions to do things conventional computers cannot handle, such as simulating the quantum processes behind chemical reactions. Most computers today work by manipulating binary digits – bits, which can each be either 1 or 0. A quantum computer instead works with a bit in a superposition of states. These quantum bits or 'qubits' each have a probability of potentially being 1 or 0 but are, in effect, both, until a measurement is made. A qubit, however, really comes into its own when it is entangled with others – ideally, with hundreds of others rather than the few dozen or so that have been corralled together to date. Physicists are probing the limits of quantum theory with experiments to find out whether such large assemblies of particles (which might be atoms, photons or electrons, for example) can be entangled and kept in quantum states long enough to do useful computing.

One result of the past few decades of experiments and theorizing is that most physicists now believe that there is no instantaneous wave function collapse. Instead, the wave function 'decays' in a small but finite amount of time into a classical state through a process called 'decoherence'. How quickly this happens depends on how isolated the quantum system is from the environment and its size.

But there are other points of view. 'Spontaneous collapse', for example, posits that wave function collapse occurs on a time scale that is inversely related to the size of the object in question. The wave function of a particle such as an electron may not collapse for 100 million years or more, but a cat's would collapse almost instantly. This solution to the measurement problem was put forward by Giancarlo Ghirardi, Alberto Rimini and Tullio Weber in 1986.[78]

Which of these many interpretations, if any, will turn out to be true? Von Neumann remained open-minded about the possibility of a deeper alternative to quantum theory for the rest of his life. 'In spite of the fact that quantum mechanics agrees well with experiment,' he says in his book, 'one can never say of the theory that it has been proved by experience, but only that it [is] the best known summarization of experience.' He was, however, more circumspect about the prospect of a future theory restoring causality. That events appear to be linked to each other in the familiar everyday world is irrelevant, von Neumann argued, because what we see is the average outcome of countless quantum interactions. If causality exists, then it needs to be found in the atomic realm. Unfortunately, the theory that best accounts for observations there appears to be in contradiction with it.

'To be sure,' von Neumann continues, 'we are dealing with an age-old way of thinking that has been embraced by all mankind. But that way of thinking does not arise from logical necessity (else it would not have been possible to build a statistical theory), and anyone who enters the subject without preconceived notions has no reason to adhere to that way of thinking. Under such circumstances, is it reasonable to sacrifice a reasonable physical theory for the sake of an unsupported idea?'[79]

Dirac, on the other hand, felt that quantum theory was not the whole story. 'I think,' he told his audience during a lecture tour of Australia and New Zealand in 1975, 'that it is quite likely that at some future time we may get an improved quantum mechanics in which there will be a return to determinism and which will, therefore, justify the Einstein point of view.'[80]

Today, we know that Dirac was almost certainly wrong, and the hopes of Einstein were misplaced. There may yet be a better theory than quantum mechanics, but thanks to Bell's work and the experiments that followed, we know that non-locality will be part and parcel of it. Conversely, von Neumann's cautious conservatism appears with hindsight the correct attitude. There is no proof yet for a deeper alternative to the quantum theory that von Neumann helped to forge more than a hundred years ago. All the experiments to date have revealed no hidden variables, nothing to suggest causality reasserts itself at some deeper level. As far as we know, it's quantum all the way down.

Physicists now doff their hats to von Neumann's Hilbert space theory, but Dirac's approach is the one most often taught to undergraduates.[81] Yet von Neumann's formulation of quantum mechanics remains definitive. Wigner, who received a Nobel for his work in quantum mechanics, insisted that his friend Jancsi was the only person who understood the theory. While Dirac laid down many of the tools of modern quantum physics, von Neumann laid down a gauntlet. He presented the theory as coherently and lucidly as anyone could and by doing so exposed the limits of quantum mechanics to scrutiny. Without a clear view of those limits, interpreting the theory is impossible. 'The historically most influential and hence for the history of the interpretations most important formalism' was that of von Neumann, says Jammer.[82] For physicists not content to shut up and calculate, von Neumann's book remains required reading today, nearly a hundred years after its publication.

Von Neumann's contributions to quantum theory did not end with his book. He helped Wigner with work that would win his friend a share of the Nobel Prize. While developing the maths of quantum theory, he became fascinated with the properties of operators in Hilbert space.[83] Operators can, for example, be added, subtracted and multiplied and so are said to 'form an algebra'. Operators connected to each other by similar algebraic relationships are dubbed 'rings'.

For several years, von Neumann outlined the properties of these operator algebras and published what he found in seven monumental papers, a total of 500 pages in length – his most profound contribution to pure mathematics. He discovered three irreducible types of operator ring that he called 'factors'. Type I factors exist in n-dimensional space, where n can be any whole number up to infinity. Von Neumann's version of quantum mechanics was expressed in just this kind of infinite-dimensional Hilbert space. Type II factors are not restricted to a Hilbert space with a whole number of dimensions; they can occupy a fractional number of dimensions, ½ or π (don't even try to visualize this). Type III factors are those that do not fit into the other two categories. The three together are now known as von Neumann algebras.

'Exploring the ocean of rings of operators, he found new continents that he had no time to survey in detail,' writes Dyson. 'He intended one day to publish a grand synthesis of his work on rings of operators. The grand synthesis remains an unwritten masterpiece, like the eighth symphony of Sibelius.'[84]

Others have since explored a few of the archipelagos and peninsulas of von Neumann's operator theory and returned with enormous riches. The mathematician Vaughan Jones, for example, was awarded the Fields Medal in 1990 for his work on the mathematics of knots, which emerged from his study of Type II von Neumann algebras. Jones had read *Mathematical Foundations* as an undergraduate. 'His legacy is quite extraordinary,' Jones says. A central aim of knot theory is to distinguish with certainty whether two tangles of string are genuinely different or if one can be turned into the other without cutting the string. Different forms of essentially the same knot are described by the same polynomial. Jones discovered a new polynomial that could distinguish between, for example, a square knot and a granny knot. The Jones polynomial now crops up in different areas of science. Molecular biologists, for instance, have used it to understand how cells uncoil the tightly knotted DNA inside the nucleus so that it can be read or copied.

Physicist Carlo Rovelli and mathematician Alain Connes, meanwhile, have used Type III factors in their effort to solve the 'problem of time': that though we feel time to flow 'forwards', there is no single unified explanation for why this is so (quantum theory and general relativity, for example, have radically different concepts of time).[85] The pair speculate that the non-commutativity at the heart of quantum theory and embedded in Type III algebras may give time a 'direction' because two quantum interactions must occur in sequence, not simultaneously. This, they claim, determines an order of events that we perceive as the passing of time. If they are right, our perception of time itself is rooted in von Neumann's maths.

The dark political clouds that had been gathering on the horizon quickly rolled in after von Neumann and Wigner left Germany in 1930. In September, the Nazi Party garnered more than 6 million votes to become the second-largest party in the Reichstag. At the next election, two years later, they received 13.7 million votes, and Hitler was appointed Chancellor of Germany in January 1933. When a fire gutted the Reichstag the following month, Hitler was awarded emergency powers. Freedom of speech, freedom of the press and the right to protest were suspended along with most other civil liberties. In March, he consolidated his power with the Enabling Act, which effectively allowed Hitler and his cabinet to bypass parliament. One of the first acts of the new regime was to introduce the 'Law for the Restoration of the Professional Civil Service', which called for the removal of Jewish employees and anyone with communist leanings. In Germany, university staff are officially appointed and paid directly by the government. About 5 per cent of all civil servants lost their jobs. But physics and mathematics departments were devastated: 15 per cent of physicists and 18.7 per cent of mathematicians were dismissed. Some lost more than half their faculty more or less overnight. Twenty of the ousted researchers were either already Nobel laureates or future recipients of the prize. Some 80 per cent were Jews.

Back in Princeton, Wigner faced a quandary. Princeton had extended his contract for five years along with von Neumann's but he felt guilty about turning his back on Europe. He turned to his friend for advice. 'Von Neumann,' Wigner said, 'asked me a simple question: Why should we stay in a part of the world where we are no longer welcome? I thought about that for weeks and came up with no good answer.' Instead, Wigner focused his efforts on finding jobs for the scientists now desperate to leave Germany.

In June that same year, von Neumann wrote to Veblen: 'If these boys continue for only two more years (which is unfortunately very probable), they will ruin German science for a generation – at least.'[86] How right he was. By the end of 1933, Germany was a totalitarian dictatorship, and the trickle of scientists leaving the country became a flood. The economist Fabian Waldinger recently analysed the impact of the dismissals on German research.[87] Scientific productivity dropped like a stone: researchers produced a third fewer papers than before. The 'Aryan' scientists recruited to replace those forced to leave were generally of a lower calibre. He found that university science departments that were bombed during the war recovered by the 1960s, but those that had lost staff remained sub-par well into the 1980s. 'These calculations suggest that the dismissal of scientists in Nazi Germany contributed about nine times more to the decline of German science than physical destruction during WWII,' Waldinger notes. By coincidence, his analysis indicated that the most influential scientists between 1920 and 1985, as measured by how often their research was cited by others, were Wigner among the physicists and von Neumann in maths.

In Göttingen, Born, Noether and Richard Courant, Hilbert's de facto deputy, were among those who left as the mathematics and physics departments were decimated. Virtually all the founders of quantum mechanics emigrated en masse. Heisenberg stayed, only to be branded a 'white Jew' for his adherence to the theories of Einstein. Hilbert surveyed the scene with utter bewilderment. He hated chauvinism. Five years earlier, Germany had been invited to its first major international mathematics conference since the end of the First World War. Many of his colleagues tried to whip up a boycott, to protest their earlier exclusion. Ignoring them, Hilbert triumphantly led a delegation of sixty-seven mathematicians to the congress. 'It is a complete misunderstanding of our science to construct differences according to people and races, and the reasons for which this has been done are very shabby ones,' he declared. 'Mathematics knows no races. For mathematics, the whole cultural world is a single country.'

As the sacked professors departed, the seventy-one-year-old mathematician accompanied them to the train station and told them their exile could not last long. 'I am writing to the minister to tell him what the foolish authorities have done.' The minister in question was, unfortunately, Bernhard Rust, who was instrumental in initiating the purges. Next year, when Rust attended a banquet at Göttingen, he asked Hilbert whether it was true that mathematics had suffered after the removal of Jews. 'Suffered?' replied Hilbert. 'It hasn't suffered, Herr Minister. It just doesn't exist anymore.'[88] Hilbert would die of natural causes a decade later in wartime Germany.

The golden age of German science was over. America was about to get an injection of talent that would transform its fortunes for ever. Von Neumann would soon be reunited with many of his Göttingen colleagues – not to discuss the finer points of quantum mechanics this time, but to design the most powerful bomb ever made.

4

Project Y and the Super

From Trinity to Ivy Mike

> *Bohr: I don't think anyone has yet discovered a way you can use theoretical physics to kill people.*
>
> *Michael Frayn,* Copenhagen, *1998*

John von Neumann watched the unfolding disaster in Europe with horrified fascination from his berth in Princeton. He had arrived newly married to his childhood sweetheart, Mariette Kövesi. They had first met in Budapest, at his brother Michael's birthday party in 1911. Mariette would share Johnny's taste for fast cars (and a penchant for driving dangerously) in later life but her chosen mode of transport that day was a tricycle. She was two and a half. The pair stayed in touch. Johnny began to court Mariette, rather bumblingly, in 1927, when she was studying economics at the University of Budapest. She was a fashionable socialite, and he was already a world-renowned mathematician, but she was never intimidated by his intellect. Von Neumann proposed, in his way, two years later. 'You and I could have a lot of fun together,' he told her, 'for instance, you like to drink wine and so do I.'

The two were married on New Year's Day 1930 and sailed to America on a luxury liner. Mariette, horribly sea sick, was confined to her quarters for the whole voyage. House-hunting did not go well at first either. Despite the generous salary Princeton was paying, nothing within their budget could match the European grandeur to which they were accustomed. 'How could I do good mathematics in a place like this?' wailed von Neumann when they went to view some of the

John von Neumann and family at breakfast in Budapest in the early 1930s after the marriage of his cousin Lili. From left to right are von Neumann, the newlyweds and Mariette Kövesi von Neumann. *Courtesy of Marina von Neumann Whitman.*

places for rent. They finally settled on an apartment that, while not as magnificent as their respective Budapest homes, was at least furnished in fittingly bourgeois European style. There were no cafés for mathematicians to congregate, so Mariette entertained von Neumann's colleagues at home.

Next, there was the problem of getting about. Von Neumann enjoyed driving very much but had never passed a test. At Mariette's suggestion, he bribed a driving examiner. This did nothing to improve his driving. He sped along crowded roads as if they were many-body problems to be negotiated by calculating the best route through on the fly. He often failed, and an intersection in Princeton was soon christened 'von Neumann corner' on account of the many accidents he had there. Bored on open roads, he slowed down. When conversation faltered, he would sing; swaying and rocking the steering wheel from side to side with him. The couple would buy a new car every year, usually

Princeton in the 1930s. Left to right: Angela (Turinsky) Robertson, Mariette (Kövesi) von Neumann, Eugene Wigner, Mary Wheeler, John von Neumann, and, on the floor, Howard Percy ('Bob') Robertson. *Courtesy of Marina von Neumann Whitman.*

because von Neumann had totalled the previous one. His vehicle of choice was a Cadillac, 'because', he explained whenever anyone asked, 'no one would sell me a tank'. Miraculously, he escaped largely unscathed from these smash-ups, often returning with the unlikeliest of explanations. 'I was proceeding down the road,' begins one fabulous excuse. 'The trees on the right were passing me in orderly fashion at 60 miles an hour. Suddenly one of them stepped in my path. Boom!'[1]

The only worse driver in Princeton was Wigner. Over-cautious rather than reckless, Wigner would roll along as far to the right as he could manage, sometimes causing pedestrians to scatter by mounting the kerb. A Princeton graduate student named Horner Kuper was asked to teach Wigner how to drive properly. Mariette would leave Johnny for Kuper in 1937.[2]

On 28 January 1933, two days before Hitler became chancellor of

Germany, John von Neumann became a beneficiary of Veblen's largesse with other people's money. Veblen had long dreamed of an independent mathematics institute where scholars on huge salaries could think big thoughts, unburdened from any irksome teaching commitments. He had shared his vision with Abraham Flexner, an influential expert on higher education who helped to secure the Rockefeller Foundation's cash to help build Princeton's Fine Hall. Flexner returned to him with more good news. The German Jewish owners of Bamberger's, a chain of department stores, had sold up to R. H. Macy & Co. and wanted to spend some of the proceeds on a new school for higher learning.

For its first six years, the Institute for Advanced Study (IAS) was housed in Fine Hall. Flexner was appointed the institute's first director in May 1930 with a salary of nearly $400,000 in modern terms. He hired Veblen as its first professor in 1932. The fifty-four-year-old Einstein joined the faculty the following year after Flexner spent much of the previous one vigorously courting him. So too did von Neumann and his fellow mathematicians Hermann Weyl and James Alexander.

Von Neumann was twenty-nine when he moved to the IAS from Princeton, the youngest of the new recruits. There was no obligation to teach (indeed there were no students), and staff were required to be on the premises for only half the year. The remainder of the time, they were officially on leave. Senior professors like Einstein earned $16,000, while an 'ordinary' professor like von Neumann received an annual salary of $10,000, close to $200,000 today. A pretty sum now and in Depression-era America almost obscene. Envious Princetonians dubbed the moneyed academy a stone's throw away 'the Institute for Advanced Salaries' and 'the Institute for Advanced Lunch'. Unused to the gilded existence, some professors would lapse into somnolence. 'These poor bastards could now sit and think clearly all by themselves, OK?' wrote one critic, the Nobel Prize-winning physicist Richard Feynman. 'They have every opportunity to do something, and they're not getting any ideas . . . Nothing happens because there's not enough real activity and challenge: You're not in contact with the experimental guys. You don't have to think how to answer questions from the students. Nothing!'[3]

Einstein would spend his Princeton years in unfruitful attempts to merge his theory of gravity with the laws of electromagnetism. Von Neumann, however, was never troubled by a lack of ideas. If anything, he had too many: one criticism of his 1930s work was that, after a few dazzling papers on a subject, he would lose interest and leave the mundane job of following them up to others. Nonetheless, his time at the IAS was to be extraordinarily productive.

One of his most impressive contributions of the period is a proof of the ergodic hypothesis. The term 'ergodic', a fusion of the Greek words *ergon*, meaning 'work' and *odos*, meaning 'way', was coined by the Austrian physicist Ludwig Boltzmann in the 1870s. Boltzmann derived the properties of a gas (such as its temperature or pressure) from the motion of particles (atoms or molecules) comprising them. Boltzmann's 'kinetic theory of gases' assumes gasses are ergodic systems: very loosely speaking, that any particular property of the gas averaged over time will be equal to its average over space. This means that whether you choose to measure the pressure of a gas inside a balloon over a long period of time or add up the pressure exerted by its atoms at any particular moment, the answer will be the same.[4]

Boltzmann did not prove his conjecture. Von Neumann did in the 1930s. But his theorem was not published first. Hearing about his work, the eminent American mathematician George Birkhoff built on his proof to create a more robust theorem. The two men met during a dinner at the Harvard Faculty Club, where Birkhoff, twenty years von Neumann's senior, refused his request to delay and rushed his work into print.[5] Von Neumann's genteel, European sense of propriety was rather offended by Birkhoff's cut-throat competitiveness. But he did not bear a grudge and afterwards became close friends with Birkhoff's son, Garrett. They wrote a paper together showing that the distributive law of classical logic, A and (B or C) is the same as (A and B) or (A and C), does not hold in the quantum world – a rather counter-intuitive result that is a consequence of the uncertainty principle. Thirty years later, the younger Birkhoff would summarize some of von Neumann's work in the 1930s. 'Anyone wishing to get an unforgettable impression of the razor edge of von Neumann's mind,' he wrote, 'need merely try to pursue this chain of exact reasoning for himself, realizing that often five pages of it were written down before breakfast, seated at a living room writing-table in a bathrobe.'[6]

This was around the time an unkempt young mathematician, eight years his junior, came to von Neumann's attention. Alan Turing's first paper, published in April 1935, developed work in von Neumann's fifty-second, on group theory, which had appeared the previous year. Coincidentally, this was exactly when von Neumann arrived in Cambridge, England, where Turing had a fellowship at King's College, to deliver a series of lectures on the same topic. The pair almost certainly met in person for the first time. They would renew their acquaintance in September the following year, when Turing arrived in Princeton as a visiting fellow. He had asked von Neumann to write him a letter of recommendation. Five days later, working in an office in Fine Hall, Turing received the proofs for 'On Computable Numbers, with an Application to the *Entscheidungsproblem*', the paper that laid the theoretical foundations of modern computer science.[7] He was disappointed by its reception in Princeton. But one person did notice it. 'Turing's office was right near von Neumann's, and von Neumann was very interested in that kind of thing,' says Herman Goldstine, who would work closely with von Neumann on computers. 'I'm sure that von Neumann understood the significance of Turing's work when the time came.'[8] Turing stayed in Princeton until July 1938. Von Neumann offered him a position as his assistant with a handsome salary attached, but Turing turned him down. He had work to do back in England.

Von Neumann's wandering interests during the 1930s may, in part, be explained by the fact that he was preoccupied with the war that he knew was coming. Three years before the outbreak of war, 'he had a rather clear picture of the catastrophes to come,' says Ulam. 'Believing that the French army was strong, I asked, "What about France?" "Oh! France won't matter," he replied. It was really very prophetic.'[9]

Von Neumann's remarkable foresight is evident in letters he wrote to Ortvay between 1928 and 1939. 'There will be a war in Europe in the next decade,' he told the Hungarian physicist in 1935, further predicting that America would enter the war 'if England is in trouble'. He feared that during that war, European Jews would suffer a genocide as the Armenians had under the Ottoman Empire. In 1940, he predicted that Britain would be able to hold a German invasion at bay (far from obvious at the time), and that America would join the war the following year (as it did after the bombing of Pearl Harbor). Von Neumann would soon be keen to turn his talents to helping his adoptive country prepare for what he felt was inevitable. In the interim, he lobbied anyone with the power to influence the course of events for America to join the war. 'The present war against Hitlerism is not a foreign war, since the principles for which it is being fought are common to all civilized mankind,' he wrote to his congressman in September 1941, 'and since even a compromise with Hitler would mean the greatest peril to the future of the United States.'[10]

Events conspired to distract von Neumann further from his work: in 1935, he became a father. Marina, von Neumann's only child, was born on 6 March. By this time, the mathematician was no longer the slim young man of the 1920s. 'He was instead rather plump, though not as corpulent as he was to become later,' remembers Ulam, who first met von Neumann in Warsaw that year. 'The first thing that struck me about him were his eyes – brown, large, vivacious, and full

John von Neumann. *Stanislaw Ulam papers, American Philosophical Society.*

of expression. His head was impressively large. He had a sort of waddling walk.'

Von Neumann offered Ulam a stipend to come to the IAS for a few months. 'He was ill at ease with people who were self-made or came from modest backgrounds,' says Ulam of von Neumann in Princeton. 'He felt most comfortable with third- or fourth-generation wealthy Jews.' Von Neumann's home life was difficult, Ulam noted. 'He may not have been an easy person to live with – in the sense that he did not devote enough time to ordinary family affairs,' adding that 'he probably could not be a very attentive, "normal" husband.'

His daughter Marina agrees. 'Although he genuinely adored my mother, my father's first love in life was thinking, a pursuit that occupied most of his waking hours, and, like many geniuses, he tended to be oblivious to the emotional needs of those around him,' she says. 'My mother, accustomed to being the centre of attention, didn't like playing second fiddle to anyone or anything, even when the competition was her spouse's supercreative mind.'[11]

What was obvious to Ulam and Marina was not to von Neumann. He was bewildered by Mariette's decision to leave him in 1937 and, according to his daughter, remained so for the rest of his life. Glum and lonely without his wife or baby daughter, von Neumann became an American citizen that year and threw himself into preparing America for the impending war.

During the First World War, Oswald Veblen had served as a captain then as major in the Ordnance Department of the US Army, charged with overseeing technical work at the newly established Ballistics Research Laboratory (BRL) of the Aberdeen Proving Ground in Maryland. The principal purpose of the lab was to study the trajectories of projectiles in flight, in an effort to improve their range and destructiveness. Guns used by the Central and Allied powers routinely lobbed shells thousands of feet in the air and over distances of several miles. Germany's infamous Paris gun had a staggering firing range of over 70 miles. But a projectile hurled high and long in this way flies through progressively thinner air as it gains altitude, and so experiences less resistance to its motion. A failure to adequately account for this meant that early efforts to calculate trajectories were wildly off, and shells flew far beyond their intended targets. Throw in some more complications – a moving target, boggy ground and so forth – and the equations of motion often become impossible to solve exactly (in mathematical terms they become 'non-linear'), forcing mathematicians to approximate. That required arithmetic and lots of it: hundreds of multiplications for a single trajectory. What was needed, but not available (yet), was a device able to perform such calculations accurately at the rate of thousands per second. Some of the earliest room-sized computers would be built to solve exactly this problem.[12]

In the run-up to the Second World War, the interests of the military turned from bigger guns to bigger bombs, delivered by planes, missiles or torpedoes. The maths of the expanding shock wave produced by a bomb blast is essentially the same as that of a shell fired at supersonic speed. Maximizing the destructive power of bombs required a detailed understanding of the hydrodynamics of explosions and the non-linear equations that accompanied them. It was to this ticklish problem – literally how to get the biggest bang for the Army's buck – that von Neumann directed his attention in the early 1930s. He would do so in an official capacity from 1937 after Veblen, recalled to his role at the Aberdeen Proving Ground, took him on as a consultant.

Hoping to make a bigger contribution to the war effort than this part-time role allowed, von Neumann applied for a commission as a lieutenant in the ordnance department. Not only would he be able to play a full role in the coming conflict, he reasoned, but a position in the Army would give him access to ballistics data that would be challenging for a civilian to obtain. Things did not go to plan. He sat the exams over the course of the next two years and, predictably, aced them, getting 100 per cent on all but one paper on military discipline. Asked what charge should be brought against a man who had deserted under complex circumstances, he advised the lesser charge of absence without leave rather than desertion. Von Neumann scored only 75 per cent because of his leniency, but his combined performance was more than good enough for him to be taken on. His chances of joining the Army, however, were scuppered when he had to defer his final exam until January 1939. He passed the exam but was turned down by a pedantic clerk who noticed he was now a couple of weeks over the age limit of thirty-five. What had cost von Neumann his post was that in September the previous year, he had left the United States for Europe to marry his second wife, Klára Dán.

Klára Dán, known to friends and family as 'Klári', first met von Neumann on the Riviera in Monte Carlo in the early 1930s. Klári was there with her first husband, Ferenc Engel, an inveterate gambler. 'When we walked into the Casino, the first person we saw was Johnny; he was seated at one of the more modestly priced roulette tables with a large piece of paper and a not-too-large mound of chips before him,' she remembers. 'He had a "system" and was delighted to explain it to us: this "system" was, of course, not foolproof, but it did involve lengthy and complicated probability calculations which even made allowance for the wheel not being "true" (which means in simple terms that it might be rigged).'[13]

Her husband eventually drifted away to another table, and Klári ordered another drink from the bar. Her marriage, 'an absolute disaster', would soon be over. In the midst of the hedonists and high rollers,

Klára (Klári) von Neumann in 1939, as pictured on her French driver's license. *Courtesy of Marina von Neumann Whitman.*

the rich and the famous, she was bored. When von Neumann joined her at the bar, she was delighted. He had run out of money. She bought him a drink.

The Dáns, like the von Neumanns, were a wealthy Jewish Budapest family who shared a magnificent house partitioned into apartments. The house was the venue for enormous gatherings of the city's elite, where businessmen and politicians rubbed shoulders with artists and writers. Even quiet family dinners would often erupt into festivities that would last all night.

'There was a bottle of wine and the confab started,' she says. 'As often as not, another bottle was passed around; pretty soon a gypsy band was summoned, perhaps some close friends cajoled out of bed and a full-fledged "mulatsag" was on its way.'

A *mulatsag*, Klári explains was 'simply the spontaneous combustion of a bunch of people having a good time'. 'At six o'clock in the morning,' she continues, 'the band was dismissed, we went back upstairs, had a quick shower, the men went to work, the children to school, and the ladies with their cooks to the market.' Klári would rekindle the spirit of those parties whenever she could in America, many years later.

Klári met von Neumann again in 1937 when he was in Budapest on a summer visit. Klári had remarried, this time to Andor Rapoch, a banker eighteen years older than she was. Von Neumann was nearing the end of his first marriage: his divorce from Mariette would come through the next month.

'We struck up a telephone acquaintance,' says Klári, 'which soon turned into sitting in cafés and talking for hours, just talking and talking. The subject of conversation veered wildly: politics and ancient history to 'the differences between America and Europe; the advantage of having a small Pekinese or a Great Dane'.

When von Neumann left on 17 August, their exchanges continued through letters and telegrams. 'It became perfectly clear that we were just made for each other,' Klári recollects. 'I told my kind and understanding daddy-husband quite frankly that nothing that he or anybody would do could be a substitution for Johnny's brains.'

Rapoch seems to have acceded to a divorce without much animus, but a mess of red tape prevented the happy couple from being wed.

Court hearings were repeatedly delayed. The Hungarian authorities refused to recognize von Neumann's divorce, while the American authorities required him to formally renounce his Hungarian citizenship before they would grant Klári a visa.

Only late the following year, with Europe on the brink of war, did Klári's divorce seem close to being finalized. Von Neumann arrived in Copenhagen in September and was staying with the Bohrs. 'It's all like a dream, a dream of peculiarly mad quality,' he wrote to Klári. 'The Bohrs quarrelling, whether Tcheckoslovakia [*sic*] ought to give in – and whether there is any hope for causality in quantum theory.'[14]

When France and Britain signed the Munich Agreement, allowing Nazi Germany to annex the Sudetenland from Czechoslovakia, von Neumann swooped in to Budapest to whisk his fiancée away to America. 'I can only say that Mr Chamberlain obviously wanted to do me a great personal favour,' he wrote to Veblen days after the agreement was concluded. 'I needed a postponement of the next world war very badly.'[15]

Klári's divorce came through at the end of October, and she married von Neumann two weeks later. They sailed to America the next month. 'I have always felt certain that my father married her on the rebound,' Marina says, 'both to assuage the hurt caused by Mariette's desertion and to provide himself with a helpmeet who could manage the everyday details of life that eluded him.'[16] The marriage would nonetheless last until von Neumann's death.

In Princeton, the couple settled into one of the grandest addresses in town. The white clapboard house at 26 Westcott Road would soon, with Klári's help, become the setting for lavish parties that were soon the stuff of Princeton legend. Alcohol flowed freely, and von Neumann, who found chaos and noise conducive to good mathematics, would sometimes disappear upstairs, cocktail in hand, to scribble down some elegant proof or other.

With the start of war now perilously close, Klári returned to Hungary to convince von Neumann's family and her own that they too should leave Europe (France and Britain's earlier capitulation had convinced them that the threat of a military conflict had been removed). Still fighting for his commission to the Army and increasingly busy with military work, von Neumann could only worry from afar. Klári's parents and Johnny's mother and brother, Michael, were safely dispatched (Nicholas appears to have emigrated earlier) but Klári stayed behind to resolve some family affairs. 'For God's sake do not go to Pest,' von Neumann begged on 10 August, 'and get out of Europe by the beginning of Sept! I mean it!'[17]

Klári sailed from Southampton aboard the SS *Champlain* on 30 August.[18] Germany invaded Poland the next day. She arrived safely, but tragedy struck a few months later. Her father, a man of wealth and influence in Hungary, was unable to adjust to his new life. A week before his first American Christmas, he stepped in front of a train. His death plunged Klári into a black depression.

Her relationship with von Neumann was not always a consolation. He was sometimes distant and aloof; 'some people, especially women, found him lacking in curiosity about subjective feelings and perhaps deficient in emotional development,' says Ulam. Klári's husband exuded nervous energy and showed signs of obsessive-compulsive disorder. 'A drawer could not be opened unless it was pushed in and out seven times,' Klári notes, 'the same with a light switch, which also had to be flipped seven times before you could let it stay.'

Veblen heard von Neumann had been denied a commission and he came to his rescue again, appointing him to the BRL's scientific advisory board in September 1940. When von Neumann showed his usefulness, other appointments quickly followed. By December he was also the chief ballistics consultant of the War Preparedness Committee of the American Mathematical Society and the Mathematical Association of America. For twelve months from September 1941, he was a member of the National Defence Research Committee (NDRC), which together with its successor, the Office of Scientific Research and Development (OSRD) would coordinate nearly all scientific research related to war. The government agency was conceived and led by Vannevar Bush, an influential engineer. Under Bush, who reported directly to the president, the NDRC and, later, the OSRD, would support a range of projects including radar, guided missiles and the proximity fuze. The most important work the agency would support would be research on an atom bomb.

The extraordinary news that uranium could be split arrived in the US by boat on 16 January 1939. German chemist Otto Hahn and his assistant Fritz Strassmann had discovered that the uranium nucleus, bombarded with neutrons, appeared to burst into fragments including barium – an element more than half its size. But with no known mode of radioactive decay to account for their observations, the chemists were baffled. It was Lise Meitner, Hahn's erstwhile collaborator, and her nephew, Otto Frisch, who first explained the physics behind their results, a type of nuclear reaction that Frisch later named 'fission'. Meitner, who had converted to Christianity but was born into a Jewish family in Vienna, had been smuggled out of Nazi Germany just a few months earlier in July 1938.[19] She had arrived in Sweden shortly afterwards, to work for a modest stipend at the Nobel Institute in Stockholm.

Informed of Hahn and Strassman's experiments over Christmas, Meitner and Frisch realized that the uranium nucleus was not rigid but unstable and wobbly, like jelly. Rather than chip off a small piece, they concluded that the impact of a neutron would force the nucleus to fly apart, releasing a huge amount of energy in the process. In the New Year, Frisch told Bohr, who was sailing for the United States the next day. Bohr agreed to keep their theory secret until their paper had been accepted by a journal but could not resist working through the details of fission aboard the boat with fellow physicist Léon Rosenfeld. Rosenfeld, blissfully unaware of Bohr's promise, travelled up by train to Princeton on the day the ship docked and gave a talk about the findings that very night. Word spread spectacularly fast. 'What do you think about the Uranium → Barium disintegration?' von Neumann asked his friend Ortvay excitedly in a letter dated 2 February 1939. 'Around here this is thought to be of great importance.'[20]

At Columbia University, Nobel laureate Enrico Fermi heard the news. He had fled fascist Italy with his Jewish wife, Laura Capon, straight after collecting the prize in Stockholm a couple of months earlier. He instantly understood the ramifications of the discovery. 'A little bomb like that,' he said, cupping his hands and gazing out of his office window onto Manhattan, 'and it would all disappear.' At the University of California in Berkeley, physicist Robert Oppenheimer understood too. On his office blackboard, there soon appeared a drawing of a bomb.

*

Von Neumann continued to work on explosives and ballistics through the early 1940s. He became an expert on shaped charges, working out how exactly the shape of an explosive affects the force and direction of the explosion when the charge detonates. But other interests far removed from military work also flourished. In 1942, he co-authored a paper with the distinguished Indian (later American) astronomer Subrahmanyan Chandrasekhar, who had also been called to the BRL to work on shock waves. The pair analysed the fluctuations of gravitational fields due to the movement of stars in an effort to understand the behaviour of stellar clusters.[21]

Von Neumann resigned from the NDRC in September 1942 to join the Navy. 'Johnny preferred admirals to generals, because the generals drank iced water for lunch, while the admirals when ashore drank liquor,' said Leslie Simon, a director of the BRL. More likely, von Neumann thought the problems the Navy needed him for were more pressing than those of the NDRC.

After spending the remaining three months of 1942 in Washington working for the Navy, von Neumann was sent on a secret mission to England for six months. Little is known of his wartime visit to Britain. He certainly assisted the Royal Navy by successfully working out German mine-laying patterns in the Atlantic. But this was a problem he appears to have cracked rather easily. He also learned more from British scientists about explosions, as they did from him. He wrote to Veblen that researchers in England had demonstrated that the shock wave from an explosion contained more energy (and so could cause more destruction) than the 'explosive reaction alone', a fact he thought would prove 'absolutely universal with high explosives in air'. At the Nautical Almanac Office in Bath, von Neumann also saw the National Cash Register Accounting Machine in action. This was a mechanical calculator that could be modified to do much more than the simple bookkeeping that its name implied. He wrote a set of instructions for it on the train back to London that day.

How he spent the rest of his time in Britain is still something of a mystery. Given what was about to occupy von Neumann's time next, it seems likely he talked to British scientists and mathematicians involved with the atom bomb project. There is also the tantalizing possibility that he met Turing again. Both men may have been thinking about how to turn Turing's theoretical 'universal computing machine' into electronic reality. The secrecy surrounding their movements during the war means that even today there is no firm evidence of a reunion, but something clearly fired von Neumann's interests in computing during his stay. Perhaps it was the cash register or perhaps a high-octane tête-à-tête with Turing, but in a letter to Veblen on 21 May 1943, von Neumann wrote that he had 'developed an obscene interest in computational techniques'.[22] He did not have too long to pursue these interests further in Britain, as he was suddenly recalled to the United States. His expertise was urgently required for the biggest science project undertaken anywhere in the world.

In September 1939, when Hitler's forces swept into Poland, few thought a weapon based on nuclear fission could be built quickly enough to affect the course of the war that had now begun. That changed in June 1941, with the report of a panel of eminent scientists charged with looking into the matter on behalf of the British government. Headed by the physicist George P. Thomson, the son of J. J. and himself a Nobel laureate like his father, the 'MAUD' committee[23] concluded that an atom bomb could be ready as early as 1943.

The principle was reasonably well understood. Fissioning uranium atoms release two or three neutrons which collide with other uranium atoms. These too split, releasing more neutrons that split more atoms and so on – a nuclear chain reaction. If there is enough material to sustain a chain reaction, a 'critical mass', huge amounts of energy are released within a few millionths of a second. Frisch and theoretical physicist Rudolf Peierls had calculated that just a few pounds of one uranium isotope, U-235, would explode with the energy of several thousand tons of dynamite.

On the basis of the MAUD report, Britain's prime minister, Winston Churchill, quickly took the decision to launch a nuclear weapons research programme, and by December, America's bomb programme, which had been idling along, was also thrown into high gear.

The arm of the OSRD charged with nuclear work morphed into what became known as the Manhattan Project. The massive effort to build the atom bomb, codenamed Project Y, would cost the US $2 billion (more than $20 billion today) and at its height employ more than 100,000 people.[24] In September 1942, the forty-six-year-old Army engineer Leslie Groves was appointed to lead it. The very next month, Groves chose Oppenheimer to head the top-secret laboratory that would develop the bomb.

Oppenheimer was not an obvious choice. A theorist with little experience of managing a large team, he would somehow have to exert his authority over scientists, many of whom had Nobel Prizes. Worst of all, from the military's standpoint, he was a left-winger whose closest associates – his girlfriend, wife, brother and sister-in-law – had been, and perhaps still were, members of the Communist Party. Even Oppenheimer's landlady in Berkeley was a communist. These facts would be used to strip Oppenheimer of his security clearance in 1954, a public act of humiliation that effectively ended his work for the government.

None of that mattered to General Groves, who had helped build the Pentagon and had a reputation for getting things done. 'The biggest sonovabitch I've ever met in my life, but also one of the most capable individuals,' said one subordinate of him later. 'He was a big man, a heavy man but he never seemed to tire.'[25] In Oppenheimer, Groves recognized another tireless leader who would get things done. 'He's a genius,' Groves would say after the war (though he too would testify against him in the end). 'A real genius. Why, Oppenheimer knows about everything. He can talk to you about anything you bring up. Well, not exactly. He doesn't know anything about sports.'[26] Groves pushed through Oppenheimer's appointment against the advice of Army counterintelligence.

The pair agreed Project Y should be sited in a remote, desolate location and eventually settled on a spot in northern New Mexico, 40 miles from Santa Fe. The Los Alamos school, a place for toughening up pallid boys from wealthy families, was on a mesa with spectacular views over the bleak wilds surrounding it, dotted with cacti and squat pine trees. The owners were happy to sell.

By the middle of 1943, plans for two 'gun-type' weapons were well advanced at Los Alamos.[27] A simple design, such a bomb explodes when a 'bullet' of fissile material is fired into another target piece to start a nuclear chain reaction. American chemist Glenn Seaborg had discovered plutonium in December 1940.[28] Like uranium, the new element could sustain a chain reaction but was easier to purify in large quantities. The assumption was that gun-type assemblies could be made to work for a plutonium bomb, codenamed 'Thin Man', as well as for the uranium one, known as 'Little Boy'.

Oppenheimer, however, knew there was a chance that the gun mechanism would not bring together two chunks of plutonium fast enough. If the plutonium being produced for the bomb decayed much faster than uranium, the bullet and target would melt before they could form a critical mass, and there would be no explosion. As a backup, Oppenheimer threw his weight behind another method of assembling a critical mass, proposed by American experimental physicist Seth Neddermeyer, who had earlier helped discover the muon and positron but in later years would conduct experiments to investigate psychokinesis. In principle, Neddermeyer's 'implosion' bomb design was straightforward enough. Arrange high explosives around a core of plutonium, then ensure these charges explode simultaneously to compress the core, which melts: as the plutonium atoms are squished together, more of the stray neutrons in the core start splitting atoms, a chain reaction starts, and the bomb detonates.

Neddermeyer's team, however, was understaffed and played second fiddle to the gun projects. Furthermore, their early experiments were unpromising. Hollow iron tubes wrapped with explosives collapsed as they hoped into dense bars. But the pipes recovered after the experiments were twisted around themselves in unpredictable ways, showing that the explosive shock wave that had compressed them was asymmetrical. For an implosion weapon to work, the explosive front would need to compress the fissile material evenly from all sides.

When Neddermeyer presented his results to the scientists and engineers at Los Alamos, one compared the problem to blowing in a beer can 'without splattering the beer'. The twenty-four-year-old Feynman, recruited shortly after being awarded his PhD at Princeton, pithily summarized the group's reaction to the design: 'It stinks,' he said. In 1943, the implosion design was therefore regarded as a long shot at best.

Oppenheimer would earn a reputation at Los Alamos for making the right call at the right time. His next decision would prove to be inspired. 'We are in what can only be described as a desperate need of your help,' he wrote to von Neumann in July. 'We have a good many theoretical people working here, but I think that if your usual shrewdness is a guide to you about the probable nature of our problems you will see why even this staff is in some respects critically inadequate.' He invited von Neumann to 'come, if possible as a permanent, and let me assure you, honored member of our staff', adding that 'a visit will give you a better idea of this somewhat Buck Rogers project than any amount of correspondence'.[29] Von Neumann accepted.

In fact, von Neumann made his first contribution to the American atom bomb project months before he arrived at Los Alamos, the culmination of his earlier work on explosions and shock waves and the knowledge he gleaned from researchers in Britain during his visit. When he was commended for his services to the US government after the war, his work at 'Site Y' would not be mentioned – unsurprisingly as so much of it was secret. Instead, when he was awarded the Medal for Merit by President Truman the citation was for his research on the 'effective use of high explosives, which has resulted in the discovery of a new ordnance principle for offensive action, and which has already been proved to increase the efficiency of air power in the atomic bomb attacks upon Japan'.

Von Neumann was irked when newspapers reported that he had received the medal for showing that a 'miss was better than a hit'. He had actually discovered that large bombs cause far more damage over a wider area when they are detonated in the air above their target than on the ground. The principle was well known, but von Neumann showed that the effect of an airburst was much larger than previously thought, and he improved the accuracy of the calculations to determine the optimal altitude of a bomb's detonation. In a report to the US Navy that deftly combined data from experiments at the Aberdeen Proving Grounds and wind-tunnel photographs from England with theoretical physics, he wrote: 'Even for a weak shock the reflected shock can be twice as strong as it is head-on, if the angle of incidence is properly chosen! And this happens at a nearly glancing angle, when a weaker reflection would have seemed plausible!'[30]

Von Neumann arrived on the mesa in September, a couple of months after receiving Oppenheimer's letter. His ideas 'woke everybody up', recalls Charles Critchfield, a mathematical physicist and ballistics expert who worked on the gun projects.[31] First, von Neumann pointed out that Neddermeyer's implosion experiments revealed little about how a real device would actually behave. The shock wave in those experiments, he explained, could be made more symmetrical by simply increasing the explosive charge surrounding the tubes. A more sophisticated approach to testing was required. Second, von Neumann suggested a better design for the implosion device, consisting of wedge-shaped charges arranged around the plutonium. Detonated simultaneously, the charges would produce focused jets that would compress the core far more quickly than simply packing it with high explosives. He consulted Teller, who had been annoying his Los Alamos neighbours with impromptu late-night piano recitals since his arrival in March. The pair came to the conclusion that, in this configuration, an implosion device would be far more efficient than a gun-type weapon: less plutonium would be required to produce an explosion of equivalent magnitude. This was big news: purifying sufficient quantities of uranium and plutonium for bombs was a major bottleneck of the project.

On hearing about the advantages of the implosion method, Groves barracked scientists for focusing on the 'safe' gun method. Oppenheimer moved to appoint the impressive Ukranian-American chemist George Kistiakowsky to work on the 'explosive lenses' that would be required to realize the device proposed by the mathematician. Just as optical lenses bend light, these shaped charges focused shock waves. Energized by von Neumann's comments, Neddermeyer put forward a more rigorous set of experiments to establish how uniform compression of the plutonium core might be achieved. The eight-man team working on the problem slowly expanded.

Von Neumann continued to spend about a third of his time on the bomb. He was perhaps the only scientist with full knowledge of the project who was allowed to come and go from Los Alamos as he pleased. When his call to Los Alamos had come, the Army and Navy had argued his work on ballistics and shock waves was too valuable to be neglected. As a result, much of von Neumann's theoretical work on the atom bomb was done in a secure office at the National Academy of Sciences in Washington, from where he also asked for a series of small-scale blast experiments to be carried out at

Von Neumann at Los Alamos in 1949. Here with Richard Feynman (middle) and Stanisław Ulam (right). *Photograph by Nicholas Metropolis; Courtesy of Los Alamos National Laboratory.*

Woods Hole in Massachusetts. Still, von Neumann visited the secret lab in the desert for a month or two each year. He loved the camaraderie, the poker games and the glorious booze- and cigarette-fuelled discussions, which could go on until all hours.[32] The Hungarian contingent was there in force, along, now, with his good friend Stanisław Ulam. That was no coincidence: von Neumann had recruited his fellow mathematician shortly after he had joined the project himself.

In between the fun and games, von Neumann was helping to perfect the implosion bomb. The theorists assigned to the programme purchased ten electrical punch-card machines from IBM to help solve the hydrodynamic equations that would reveal whether the explosive lenses would compress the plutonium core quickly enough to detonate the bomb. Until then, the 'computers' at Los Alamos were nearly all women working with mechanical desk calculators. Groves had insisted there was no money to waste housing civilians, so many of these human computers were the wives of physicists and engineers already working on the project. But the IBM machines did not tire and had no young children to be picked up from school or nursery. The women were superseded. For now.

The spring of 1944 brought unwelcome news for the scientists racing to develop a plutonium bomb. The Jewish-Italian nuclear physicist Emilio Segrè, who had worked with Fermi in Rome, had uncovered a problem with samples of plutonium produced by reactors in Hanford, Washington and Oak Ridge, Tennessee. With three graduate students, Segrè had been studying the fission rates of plutonium batches in a log cabin some 14 miles from Los Alamos. Within days of receiving their first batch of reactor-plutonium the team found spontaneous fission rates to be five times higher than that from cyclotrons. As Oppenheimer had feared, a gun-type weapon made with plutonium would not work.[33] The Thin Man device was abandoned.

Oppenheimer's intuitive decision to call on von Neumann's expertise six months earlier now paid dividends. Although Los Alamos was sure that the relatively simple Little Boy (the uranium gun) would explode as expected, there was no guarantee the uranium payload would be produced in time. In any case, one bomb was never going to be enough to satisfy the government or military after the huge public investment that had been made. Only a plutonium bomb, along with the know-how to easily produce more, would do.

Though many at Los Alamos still thought the implosion device would be impossible to deliver in time for combat use, Groves demanded that the project change course to prioritize it. Oppenheimer complied. Two new teams, the Gadget and Explosive Divisions, were created to work on the core and lenses. The responsibilities of dozens of scientists and engineers changed overnight. The odds had shifted. Los Alamos was betting on implosion.

By July 1944, von Neumann and others had worked out the shape of the explosive lenses to be used in the implosion device. Intensive work now began on finalizing the design of the 'gadget', as the bombs were nicknamed by those on the mesa. Every part had to be tested as thoroughly as possible. The Explosive Division struggled to find the right mix of explosives that would produce shock waves that would crush the plutonium evenly. Casting explosive lenses of the perfect shape proved to be extraordinarily difficult. Over 20,000 lenses were actually used in testing during the project, but several times this number were rejected or destroyed. The detonator had to be an electric circuit that would survive the cold air at high altitude and the buffeting it would receive as the bomb fell, yet still detonate all the lenses simultaneously. Even the number of bolts that should be used to secure the two halves of the bomb's outer armoured shell was a critical question: too few, and the lens assembly would not be held together tightly enough; too many, and assembling the bomb would take too long. Yet by February 1945, a blueprint of the implosion gadget was ready – though no one at the lab was confident it would work. Assembly of the first implosion device began straightaway. There were cracks and pockets in some of the lenses. Repairs required nerves of steel: the team drilled like dentists through the brittle material and poured in molten high explosive to fill the cavity.

There were 32 lenses in all, 20 hexagonal and 12 pentagonal blocks of high explosive carefully fitted together to form a truncated icosahedron. It resembled nothing more closely than an over-sized soccer ball.[34] This outermost shell was 47 cm thick. Inside it were concentric hollow spheres of decreasing size, each nestled inside the next, each serving the same deadly purpose of maximizing the bomb's explosive power.

The first layer of this giant apocalyptic onion was 11.5 cm of aluminium called the 'pusher', designed to enhance compression of the plutonium core by preventing a steep drop in pressure behind the shock-wave front.

Next came a 120 kg shell of natural uranium (unrefined uranium, composed mostly of the non-fissile isotope uranium-238, was in plentiful supply) – the 'tamper'. Its purpose was to delay the expansion of the plutonium core inside, so allowing the chain reaction to proceed for a fraction of a second longer after detonation. For every 10 nanoseconds the tamper held the core together, another generation of neutrons would blossom inside the fissioning plutonium, violently converting more mass to energy.

A hole drilled through this tamper allowed the plutonium pit, an apple-sized 6.2 kg ball 9 cm in diameter, to be inserted into the device at the very end. This was a sub-critical mass, to be squeezed into criticality by the shock wave. The plutonium pit was itself composed of two hemispheres with a 2.5 cm cavity in the middle to hold 'urchin', the initiator, made of polonium and beryllium and designed to trigger a chain reaction in the plutonium.

Half the size of a golf ball, the initiator was a tour de force of precision engineering. The polonium isotope the scientists used, Po-210, releases alpha particles that, on striking beryllium, liberate a burst of neutrons. The two elements can be kept apart easily enough: alpha particles cannot penetrate more than a few hundredths of a millimetre into metal. But the initiator had to be carefully designed so that beryllium and polonium also mixed thoroughly the moment the plutonium core was compressed. This was achieved by plating a nugget of beryllium with nickel and gold and depositing polonium on the surface. The nugget was itself enclosed by a shell of nickel and gold-plated beryllium, which had fine grooves cut into its inside surface to hold more polonium. The implosion shock wave would crush the initiator, instantly dispersing the polonium sandwiched between the inner and outer spheres of beryllium.

The plutonium implosion gadget, weighing several thousand kilos, was assembled on 11 and 12 July. The gun-type bomb had been waiting since May for its payload. Uranium trickled in to Los Alamos, allowing castings of the bullet and target to be made in July. As a result, Little Boy was only combat ready two weeks after the first implosion bomb was made. But the first implosion bomb would not be used in war – it was for a test.

In December 1943, von Neumann had called for a small-scale test, a deliberate 'fizzle', to put his hydrodynamics calculations to the test. The device, he suggested, could be placed inside a 10-foot-wide armoured box strong enough to contain the explosion. Unfissioned plutonium could then be recovered later by washing the container's insides. Oppenheimer later ruled that only a full-scale test would provide any assurance that the complex device would detonate in battle. But the idea that the bomb be detonated within a vessel, later codenamed 'Jumbo', stuck, and a 14-inch-thick steel cylinder weighing around 200 tons was built for the job. It was never used, one concern being that if the bomb did actually produce an explosion even a fraction of the expected size, Jumbo would instantly be transformed into 200 tons of radioactive shrapnel. Groves, fearing Congress would regard the $12 million

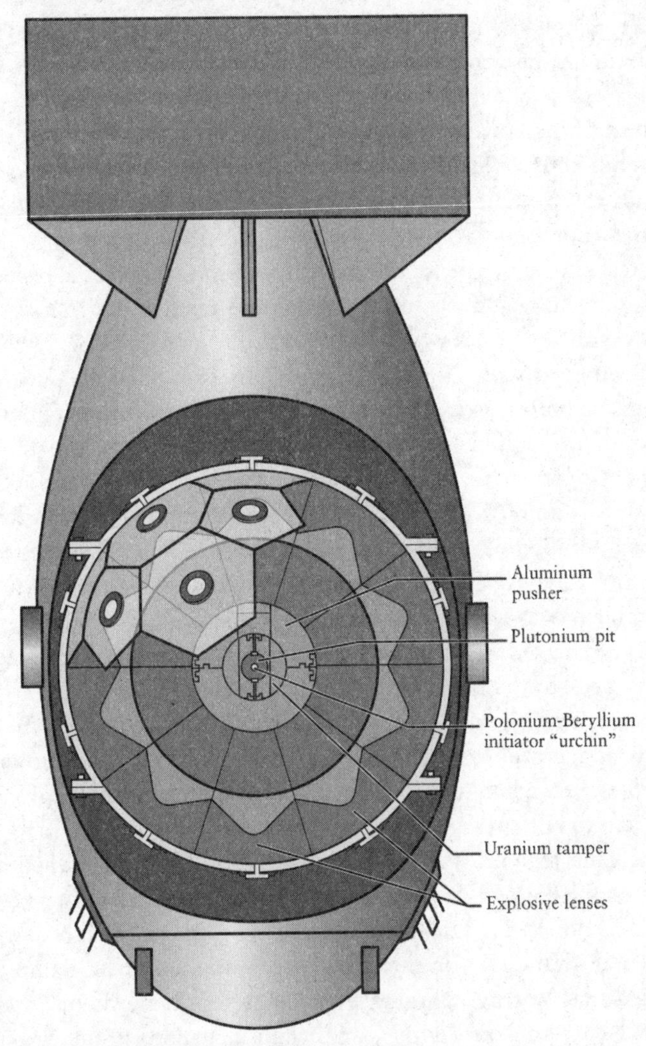

The Fat Man bomb was essentially the Trinity device encased in a protective aerodynamic steel case.

Jumbo to be a white elephant, ordered it destroyed. Several 500 pound demolition bombs could only blow the ends off the vessel, however. Jumbo's rusting hulk still stands in the New Mexico desert today.

*

Plans to detonate the world's first nuclear weapon were rapidly being finalized. The test, codenamed 'Trinity', would take place on Monday, 16 July 1945. The date had been chosen by Groves to coincide with the start of the Potsdam conference, where Truman was to meet with Churchill and the Soviet leader Joseph Stalin. Truman hoped a successful test would strengthen his hand in the discussions over the future of Europe.

A test site had to be found that was not too far from Los Alamos, so equipment could be transported there readily, and flat, so the effects of the blast wave could be observed from a distance. The committee charged with the decision chose an 18-by-24-mile swathe of land in the northwest corner of the Air Force's Alamogordo bombing range, about 200 miles south of Los Alamos.

The gadget left the mesa by truck just after midnight on 13 July. The plutonium pit with its beryllium-polonium initiator was inserted on the afternoon of 13 July. The full bomb was hoisted to the top of a 90 foot tower to simulate as well as possible the effects of an air burst. There was little left to do now but wait.

At 10 p.m. on 15 July, American physicist Kenneth Bainbridge led a team back to the base of the tower to begin the process of arming the bomb. Several hours later, the American physicist threw the last switch of the arming circuit, so that the bomb could be detonated from S-10000, a bunker 10,000 yards south of the tower. A light drizzle began to fall, presaging the thunderstorms that would shortly tear through the region. At Los Alamos, some of the staff listened anxiously to the tempest now breaking outside their cabins. A test could only be conducted in fair weather. Otherwise, the risk that heavy fallout would rain down on populated areas was too great. Bainbridge worried that a stray lightning bolt might set off the bomb prematurely. At 2 a.m. on 16 July, amid lightning and 30-mile-an-hour winds, an enraged General Groves turned on the forecaster, Jack Hubbard, who had accepted the offer to become the Trinity test's chief meteorologist just three months earlier. Hubbard reminded the fuming Groves that he had recommended the test be run either later than 18 July or earlier than 14 July. He then predicted the squall would abate by dawn, probably one of the more consequential weather forecasts in history. Promising to hang him if he was wrong,

Groves shifted the time of the test from 4 a.m. to 5.30 a.m. Luckily for Hubbard, the weather duly cleared.

At S-10000, Bainbridge authorized the start of the twenty-minute countdown at 5.09.45 a.m. A few hundred scientists and VIPs had gathered at Compañia Hill, the designated observation point 20 miles northwest of the tower. Von Neumann was among them. The scientists wagered on the explosive yield of the bomb. Some still thought zero was the most likely figure. Downplaying the chance of success, Oppenheimer chose 300 tons TNT equivalent. Teller, an optimist where bombs were concerned, picked 45,000 tons and passed around a bottle of suntan lotion. The sight of scientists slapping on sunscreen in the pitch dark perturbed some of the assembled VIPs.

At 5.29 a.m., just before sunrise, an electric pulse from 10,000 yards away detonates the thirty-two atomic lenses arrayed around the pit. The shock waves, initially convex, become concave as they travel through the carefully configured layers of explosives, merging, finally, into a rapidly contracting spherical front. The force of the implosion compresses the core to less than half its original size. The shock wave then reaches the centre of the pit, crushing the initiator and blending polonium with beryllium. In the following ten billionths of a second, nine or ten neutrons are liberated. It is enough. About a kilogram of the liquefied plutonium and some of the uranium tamper, fissions. A gram of matter is converted to energy. Then everything vaporises.

Even on Compañia Hill, 20 miles from ground zero, anyone not wearing eye protection, even those not looking directly at the explosion, was temporarily blinded by the flash. A few streaks of gold to the east had heralded the sunrise. Now a second, more menacing sun rose to challenge it. Those observing through the tinted welder's glasses provided first saw an upturned yellow hemisphere twice as wide as the sun, then a fireball that rose rapidly on a column of smoke. Their faces, cold in that desert morning, were suddenly bathed in the warmth of a hot summer's day. The ball turned red, colouring the underside of the clouds above a bright pink, before radiation ionized the air electric-blue. The illumination faded from green to white. Then it was gone. Those on the hill were blasted by the roaring shock wave about forty seconds after the flash.

'That was at least 5,000 tons and probably a lot more,' von

Neumann said quietly. Fermi had ripped a sheet of paper into pieces, letting them fall when the air blast hit. They were blown about 8 feet. Consulting a table he had prepared earlier, he declared the blast equivalent to 10,000 tons of TNT. They were both off. The best estimates of Trinity's power put the figure somewhere between 20,000 and 22,000 tons. Oppenheimer reached for poetry, recalling a verse from ancient Hindu scripture, the *Bhagavad Gita*, which he had read in the original Sanskrit. 'Now I am become Death,' he said, 'the destroyer of worlds.' Bainbridge was pithier. 'Now we are all sons of bitches,' he told Oppenheimer.

'The war is over,' Groves' deputy, Thomas Farrell, declared at Base Camp, 5 miles south of S-10000. 'Yes,' Groves replied, 'after we drop two bombs on Japan.'

At the site was a crater 500 feet wide and 6 feet deep. Sand had fused in the heat to form a thin layer of glass, later called 'trinitite'. After the explosion, some of the glass had rained down up to 1,000 yards away as spherical globules. Most were pale green due to iron in the sand; some, containing melted fragments of the bomb tower, were black, while others were coloured red and yellow by copper in the wiring. Mildly radioactive, glass taken from the site sometimes caused burns if it was worn next to the skin as jewellery in the first few years after the test.

In Potsdam, Truman had decided quite quickly that it would not, after all, be in America's interests for the Soviet Union to enter the Pacific war. When he got word of the successful test, he deliberated over whether he could tell Stalin about the bomb without provoking a Soviet invasion of Japan. In the end, Truman sidled over to the Russian leader and told him that the United States now possessed a 'new weapon of unusual destructive force'. Stalin, completely unperturbed, told him to make good use of it. Spies in the Manhattan Project had long ago told Stalin everything he needed to know.

On the morning of the Trinity test, Little Boy had set sail for Tinian, a Japanese island, 1,500 miles from Tokyo, taken by American forces the previous summer. By 2 August, the parts for three Fat Man implosion bombs had arrived there too. One, unarmed, was dropped in the Pacific on 8 August – a final test. Another was readied for the next day. The third would never be used at all.

Szilard had campaigned successfully for the United States to build the bomb. By the time Germany surrendered in May 1945, however, he was arguing that using the bomb against civilian populations would be wrong. He organized a petition urging that the bomb only be used against the Japanese as a last resort, convincing Wigner and sixty-eight other, more junior, scientists to sign. But the decision to drop the bomb on Japan had, effectively, already been made. On 23 April 1945, Groves had written a memo to Henry Stimson, updating the secretary of war on the Manhattan Project. 'While our plan of operations is based on the more certain, more powerful, gun type bomb, it also provides for the use of the implosion type bombs as soon as they become available,' wrote Groves. 'The target is and was always expected to be Japan.'[35]

There is evidence from as early as May 1943 that Germany was not in the cross-hairs of the American government. Policy-makers feared that if an atom bomb failed to detonate on German soil, scientists there would be better placed to use it to aid their own bomb-making efforts than their Japanese counterparts. Some historians have argued that racism played a role: the Japanese were despised by the American public. The internment of thousands of Japanese-Americans during the war, for example, was later investigated and found to be based on 'racial prejudice, wartime hysteria, and a failure of political leadership' rather than any real security risk.[36] Another motive was revenge for the attack on Pearl Harbor. Whatever the reasons, had the émigré scientists working on the Manhattan Project known that the ultimate destination of the bomb they were helping to build was Imperial Japan rather than Nazi Germany, many might have quit – or never joined at all.

Von Neumann had no such qualms. His first-hand experience of Béla Kun's Hungary and what he had seen of Nazi Germany instilled in him a horror of totalitarian dictatorship. With Germany defeated, he now felt that the greatest threat to world peace was Stalin's Soviet Union. Only the speedy and determined development of nuclear arms would keep Stalin in check. Otherwise, the war might end with the Soviet invasion and occupation of Japan, giving Stalin an outpost in the Pacific. This was no idle concern. Shortly after the end of the war, Stalin asked Truman to grant his 'modest wish' to occupy Hokkaido,

Japan's second-largest island. Truman refused. Possession of the atom bomb had stiffened the resolve of the American government.

Groves assembled a committee to establish a list of possible targets for the two bombs. He and Oppenheimer were eager to have von Neumann on the panel for his cool, dispassionate approach to morally fraught matters as much as for his expertise on shock waves and explosions. After a meeting at Los Alamos on 10 and 11 May, the committee was ready to send their recommendations to Groves. The US Air Force was proceeding doggedly with its plan to level every major Japanese city by 1 January 1946. It had sent a list of targets that might be left standing by the summer: (a) Kyoto, (b) Hiroshima, (c) Yokohama, (d) the imperial palace in Tokyo, (e) Kokura arsenal and (f) Niigata. The intelligence services had also provided a list of possibilities. The inclusion of dockyards and ironworks, however, betrayed a certain naivety about the power of the new weapon. Von Neumann's notes from the May meeting show that the only target on this list he approved of was one also on the Air Force's list. He wrote 'O.K.' next to Kokura arsenal and returned to the Air Force list. He struck the imperial palace from it and called for any future decision to bomb this target to be 'referred back to us'. Niigata he also rejected, asking for 'more information' before making a decision.

Von Neumann's votes in the end were for a) Kyoto, b) Hiroshima, c) Yokohama and d) Kokura arsenal – exactly the recommendations the committee as a whole were to make later.[37] Stimson was completely opposed to bombing Kyoto, the cultural heart of Japan, for eleven centuries its capital, and a city he had visited during his honeymoon. The port of Nagasaki was chosen instead. By August, Yokohama had been bombed so thoroughly that it was removed from the list, leaving only Hiroshima, Kokura arsenal and Nagasaki.

On 6 August, the *Enola Gay* took off before dawn from Tinian. Good weather allowed Little Boy to be dropped on the primary target of Hiroshima, but a crosswind blew the bomb away from the aiming point, Aioi Bridge, so that it detonated 1,900 feet (580 metres) above Shima hospital, modelled by its founder on the Mayo clinic in Rochester, Minnesota.

The Target Committee had recommended a detonation height of

2,400 feet for a 15 kiloton bomb for maximum effect.[38] More powerful bombs do more damage when exploded at greater altitude, but no one could predict with any certainty the yields of Little Boy or Fat Man. Von Neumann's notes from the meeting show he had calculated that a detonation 40 per cent below or 14 per cent above the optimum height would reduce the area of damage by about 25 per cent. Too low was preferable to too high.

The explosion, equivalent to about 17,000 tons of TNT, and the resulting firestorm killed some 70,000 people, mostly civilians. Many thousands more would die from burns and radiation poisoning by the end of the year.

The *Enola Gay* took to the skies again three days later, this time acting as the weather reconnaissance aircraft for *Bockscar*, the B-29 bomber carrying Fat Man. The skies over Kokura, initially clear, clouded over with smoke from Yahata, firebombed the previous day. After three unsuccessful attempts to drop its payload, *Bockscar* flew on to the secondary target of Nagasaki. Here, too, *Bockscar*'s bombardier struggled to sight the target. With fuel running low, the crew dropped Fat Man through a last-minute break in the clouds into a valley in the north of the city, nearly 2 miles north of its aiming point. When the bomb detonated at an altitude of about 1,650 feet (500 metres), the people of Nagasaki were therefore partly shielded from the 21 kiloton blast by the intervening hillside. Estimates of the death toll vary from 60,000 to 80,000.

Just 800 yards from ground zero in Nagasaki, Shigeko Matsumoto was one of the lucky survivors. 'My siblings and I played in front of the bomb shelter entrance, waiting to be picked up by our grandfather,' she recalls.[39]

> Then, at 11:02am, the sky turned bright white. My siblings and I were knocked off our feet and violently slammed back into the bomb shelter. We had no idea what had happened.
>
> As we sat there shell-shocked and confused, heavily injured burn victims came stumbling into the bomb shelter en masse. Their skin had peeled off their bodies and faces and hung limply down on the ground, in ribbons. Their hair was burnt down to a few measly centimeters from the scalp. Many of the victims collapsed as soon as they reached

> the bomb shelter entrance, forming a massive pile of contorted bodies. The stench and heat were unbearable.
>
> My siblings and I were trapped in there for three days.
>
> Finally, my grandfather found us and we made our way back to our home. I will never forget the hellscape that awaited us. Half burnt bodies lay stiff on the ground, eye balls gleaming from their sockets. Cattle lay dead along the side of the road, their abdomens grotesquely large and swollen. Thousands of bodies bobbed up and down the river, bloated and purplish from soaking up the water. 'Wait! Wait!' I pleaded, as my grandfather treaded a couple paces ahead of me. I was terrified of being left behind.

Stalin had declared war on Japan earlier the same day, invading the Japanese puppet state of Manchuko on three fronts. Japan surrendered unconditionally on 15 August.

Little Boy and Fat Man claimed more lives in minutes than the senseless firebombing of Dresden by hundreds of Allied (mostly British) aircraft. No decision of such magnitude should pass into the annals of history unscrutinized, and no schoolchild taught to accept uncritically that the horrors visited on the citizens of those two cities can be justified. 'The experience of these two cities', concludes a study of Hiroshima and Nagasaki written by Japanese scientists and doctors some thirty-six years after the bombings 'was the opening chapter to the possible annihilation of mankind.'[40]

Freeman Dyson worked in the research division of Britain's Bomber Command, which from 1942 began raids targeting whole German cities. He reflected on his time there with brutal honesty in his autobiography. 'I began to look backward and ask myself how it happened that I let myself become involved in this crazy game of murder,' he says. 'Since the beginning of the war I had been retreating step by step from one moral position to another, until at the end I had no moral position at all.'[41] The story of the Second World War as a whole is, likewise, one of progressive moral retreat; the bombings of Hiroshima and Nagasaki a last hideous refutation of the idea that the lives of civilians mattered, and that any war could be 'ethical'.

*

The German atom bomb project, with Heisenberg as its scientific head, had never got off the ground. Back in June 1942, Heisenberg had spoken at a meeting attended by Albert Speer, Hitler's powerful minister of armaments and war production. With military and industrial leaders in attendance, one of Speer's deputies had asked Heisenberg to estimate the size of an atom bomb capable of levelling a city. 'As large as a pineapple,' he replied, cupping his hands as Fermi had done in his office a few years earlier. Speer was impressed and wanted to throw his weight behind the project. He offered to build cyclotrons bigger than any in America. Heisenberg, however, made only modest demands and implied that a bomb could not be built in time to affect the war's outcome. The Nazis lost interest.

Heisenberg would say after the war that moral scruples had prevented him from pushing ahead, but there is little evidence that is true. He was a German nationalist through and through. True, he was not enthusiastic about giving Hitler the bomb but he would have done his utmost to do so had he believed there was any danger that the Allies would secure one first. This was an outcome he dismissed because he thought Germany was well ahead of other nations in nuclear research. As it was. Until 1933.

'Isn't it wonderful that the war is over?' said a ten-year-old Marina in a letter to her stepmother just a few days after Japan's surrender. 'Is Daddy still going to travel so much now that the war is over? I hope not.'[42] Unfortunately for the young Marina, her father's schedule became, if anything, more frenetic. His peripatetic life placed stress on his second marriage. When he was at home, he and Klári argued, the rows tempestuous enough to shatter his otherwise implacably calm demeanour. 'I never saw him lose his temper, except maybe two or three times,' says Marina. 'Klári knew how to push him far enough so finally he would explode.'[43]

The following summer, Marina went on a road trip with her father, some small recompense for all his absences. Starting from their home in Princeton, they drove across the country in his Buick, von Neumann's substitute tank that year. 'He drove at a hundred miles an hour,' she said. 'There were just the two of us. We were together a long time and talked a lot.'[44]

John and Marina von Neumann, aged 11, in Santa Fe. *Courtesy of Marina von Neumann Whitman.*

They stayed in a series of cheap motels, many with no indoor plumbing. When they arrived in Santa Fe, he bought her two silver and turquoise belts made by one of the local Native American tribes. 'Once we reached Los Alamos, my father exuded boyish enthusiasm as he showed me around those areas I was allowed to enter – all the buildings where bomb-related work was conducted remained strictly off limits,' she says. He introduced her to his Manhattan Project colleagues, including Stan and Françoise Ulam. 'The secret city,' she continues, 'whose existence had been revealed less than a year before, looked amazingly primitive to my citified eyes, with muddy paths instead of sidewalks and open stairs leading up to second-floor apartments in flimsy wooden buildings.'[45]

Her father slept only three or four hours a night during the trip, Marina says, but miraculously they arrived at their ultimate destination of Santa Barbara, California without incident. Klári met them there, and Marina's father left on a secret trip while Marina and her stepmother enjoyed a beach holiday. 'I got the worst sunburn I ever had in my life,' she recalls.[46]

It was July 1946. Von Neumann had gone to Bikini Atoll, an idyllic coral reef in the Marshall Islands. There he would witness Operation Crossroads, the first test of a nuclear weapon since Trinity. It would not be the last.

That year, von Neumann also filed a patent with Klaus Fuchs, a gifted German theoretical physicist. Fuchs was a member of the Communist Party of Germany (KPD) and had left Germany in 1933. He became a British citizen in 1942 and joined 'Tube Alloys', Britain's secret atom bomb project, at the invitation of Rudolf Peierls. Fuchs moved to the United States with Peierls to work on the Manhattan Project the following year. From August 1944, he was in the Los Alamos Theoretical Physics Division, where he would, among other things, help design the implosion bomb. In April 1946, Fuchs and von Neumann attended a conference at Los Alamos on thermonuclear weapons. Their jointly authored patent is somewhat euphemistically entitled *Improvements in Method and Means for Utilizing Nuclear Energy*. In fact, the patent contains plans for a thermonuclear weapon.

The Fuchs-von Neumann design was a version of Teller's 'Classical Super', an attempt to imitate on Earth the physics that makes a star turn supernova. Teller's idea was to use a gun-type fission bomb to trigger a fusion reaction at one end of the weapon. The shock wave and heat from that explosion would then, he hoped, cause a runaway fusion reaction to propagate down a tube of liquid deuterium (or a mixture of deuterium and tritium which ignites at lower temperatures).[47] Teller estimated that such a bomb might release energy equivalent to about 10 megatons of TNT; about a thousand times more than Little Boy.[48] The challenge was to recreate for a split second the temperatures and pressures, equivalent to those at the heart of the Sun and many millions of times that on Earth, needed to initiate and maintain fusion all the way down the column of fuel. Fuchs and von Neumann realized that they could enlist the radiation produced by the fission bomb to help.

Detonation of a gun-type device releases a massive burst of X-rays. Their patent proposes using this salvo of radiation to heat a bulb of fusion fuel surrounded by a shell of beryllium oxide (the tamper). Both the tamper and fuel almost immediately vaporise. The hot beryllium oxide gas, they suggested, would then squeeze the fuel enough to start fusion. Compressing a target in this way using electromagnetic waves is now known as 'radiation implosion'. This was the first time anyone had thought of it.

Von Neumann and Ulam's calculations would eventually show the Classical Super configuration would never be able to achieve the temperatures and pressures required to sustain a fusion reaction. The Fuchs-von Neumann bomb based on it would have fizzled. The problem would only be solved years later with the Ulam-Teller design on which modern thermonuclear weapons are based. And one of the key aspects of that successful design would be its use of radiation implosion. When the patent was filed in 1946 the means of modelling the physical processes behind the idea were not available, says physicist German Goncharov, a member of the team that developed Russian thermonuclear weapons.[49] The proposal was 'far ahead of its time,' he adds. 'It would take another five years in the US for the enormous conceptual potential of the proposal to be fully substantiated.' By that time, Fuchs would be languishing in an English jail. For Fuchs, the committed communist, was also a committed Soviet spy. In March 1948, he had sent details of the design to Russia.

The Fuchs-von Neumann patent is still classified in the United States but its contents are now known. Starting in the 1990s, post-Soviet Russia began declassifying and publishing historical documents relating to their bomb project – including a host of Manhattan Project papers obtained through espionage. Fuchs claimed some credit for the discovery of radiation implosion in 1950 while serving a fourteen-year prison sentence for breaking the Official Secrets Act. Von Neumann, on the other hand, never spoke about it. He was probably not keen to draw attention to his top-secret work with a self-confessed Soviet spy.

The intelligence galvanized Soviet scientists. They noted with interest the new approach to ignition, heating the tamper and fuel with radiation. The chief effect of the material from Fuchs, however, was to convince the scientists to press their government to start a Super project of their own. 'The Soviet political leadership interpreted the new intelligence documents on the superbomb . . . as a sign that the USA had, possibly, made considerable progress,' Goncharov says, 'so they . . . decided to launch a comprehensive programme officially supported by the central authorities.'[50]

A year before testing their first fission bomb, and well before Truman decided to accelerate America's Super programme, Russia started briskly down the path leading to the test of its first fully fledged thermonuclear weapon on 22 November 1955. Determined to help the United States keep Stalin in check, von Neumann had inadvertently helped to kick-start the Soviet H-bomb programme.

Von Neumann's trips to Los Alamos continued long after the war was over, when he and Klári would work on Teller's 'Super'. The design for this was even more complex than the implosion weapon, requiring the explosion of a fission bomb to trigger a fusion reaction within. To carry out the calculations needed to see if their designs would work, the scientists and engineers working on the Super would need a new kind of machine that was being developed at the University of Pennsylvania's Moore School of Electrical Engineering. Von Neumann soon got involved. The first modern computer was on its way.

5

The Convoluted Birth of the Modern Computer

From ENIAC to Apple

'Computers in the future may have only 1,000 vacuum tubes and perhaps weigh only 1½ tonnes.'

Popular Mechanics, *March 1949*

One morning in the spring of 1945, when von Neumann was busily shortlisting Japanese targets for the Los Alamos bombs, he arrived home from the mesa, went straight to bed and slept for twelve hours straight. 'Nothing he could have done would have had me more worried than Johnny skipping two meals,' his wife Klári noted in her memoirs, 'not to speak of the fact that I have never known him to sleep that long in one stretch.'[1]

When he awoke late that night, von Neumann began to prophesy at great speed, stuttering as he did when he was under strain. 'What we are creating now,' he told her,

> is a monster whose influence is going to change history, provided there is any history left, yet it would be impossible not to see it through, not only for the military reasons, but it would also be unethical from the point of view of the scientists not to do what they know is feasible, no matter what terrible consequences it may have. And this is only the beginning! The energy source which is now being made available will make scientists the most hated and also the most wanted citizens of any country.

But then von Neumann abruptly switched from talking about the power of the atom to the power of machines that he thought were 'going to become not only more important but indispensable'.

'We will be able to go into space way beyond the moon if only people could keep pace with what they create,' he said. And he worried that if we did not, those same machines could be more dangerous than the bombs he was helping to build.

'While speculating about the details of future technical possibilities,' Klári continues, 'he got himself into such a dither that I finally suggested a couple of sleeping pills and a very strong drink to bring him back to the present and make him relax a little about his own predictions of inevitable doom.'[2]

Whatever the nature of the vision that possessed him that night, von Neumann decisively turned away from pure maths to focus single-mindedly on bringing the machines he feared into being. 'From here on,' Klári concludes, 'Johnny's fascination and preoccupation with the shape of things to come never ceased.'

Von Neumann's interest in computing can be traced back to the 1930s.[3] During his early work for the Army, he concluded that the calculations required to model explosions would quickly swell beyond the number-crunching abilities of contemporary desk calculators. Von Neumann predicted that 'There was going to be an advance in computing machines that would have to work partly as the brain did,' according to journalist Norman Macrae, and 'such machines would become attached to all large systems such as telecommunication systems, electricity grids and big factories'. The Internet was conceived many times over before computers were linked together in the 1960s and '70s to form the ARPANET.

Had von Neumann's interest in computing been catalysed by Turing during the war? That the two men would seek each other out seems likely. The first head of Britain's National Research Development Corporation (NRDC), scientist Tony Giffard, maintained they did.[4] 'They met and sparked one another off,' he told computer scientist and historian Brian Randell in 1971. 'Each had, as it were, half the picture in his head and the two halves came together during the course of their meeting.'[5]

Whatever happened to von Neumann in Britain in 1943 – and the trail now seems to have gone cold – he became the chief advocate for computing technologies at Los Alamos when he returned to America. In January 1944, he wrote to Warren Weaver, the head of the OSRD's Applied Mathematics Panel, asking for his help in locating the very fastest calculating machines in the United States. The mathematics of the implosion device was getting out of hand. As Ulam recalls:

> I remember discussions with von Neumann in which I would suggest proposals and plans to calculate by brute force very laboriously – step by step – involving an enormous amount of computational work, taking much more time but with more reliable results. It was at that time that von Neumann decided to utilize the new computing machines which were 'on the horizon'.[6]

Weaver referred von Neumann to Howard Aiken, a physicist at Harvard University who was expecting the arrival from IBM of an electromechanical computer he had designed. The ASCC (Automatic Sequence Controlled Calculator) would be renamed the Harvard Mark I. Von Neumann visited Aiken, then returned to Los Alamos and suggested that one of their classified problems be modified to remove any indication of its real purpose. Aiken did not know it, but one of the first set of calculations to be run on his machine was a series of shock wave simulations for the bomb. But the Harvard Mark I proved slower (though more accurate) than the punched-card machines at Los Alamos and was in any case already booked up with work for the Navy. Von Neumann continued to criss-cross America for many years in an energetic hunt for computing power.

'In all the years after the war, whenever you visited one of the installations with a modern mainframe computer, you would always find somebody doing a shock wave problem,' says Martin Schwarzschild, one of the first astronomers to use electronic computers for research. 'If you asked them how they came to be working on that, it was always von Neumann who put them onto it. So they became the footprint of von Neumann, walking across the scene of modern computers.'[7]

*

Oddly, Weaver had not told von Neumann about the electronic machine being developed at the Moore School of Electrical Engineering at the University of Pennsylvania. Nor did von Neumann's mentor, Oswald Veblen, who had approved funding for the project in April 1943. Earlier devices, such as the Mark I at Harvard or Konrad Zuse's Z3, had used the positions of cogs, relay switches and gears to represent digits. But the Moore School's ENIAC (Electronic Numerical Integrator and Computer) would have no moving parts. With vacuum tubes and only electric circuitry, its designers predicted boldly that their machine would compute thousands of times faster than its predecessors.

Perhaps neither Weaver nor Veblen was confident that the untried team developing the ENIAC would deliver. Or maybe they thought that von Neumann needed access to a machine straightaway while the ENIAC was not expected to be ready for another two years. In any case, von Neumann only learned of that machine by chance, while waiting for his train home after a meeting at the BRL (Ballistics Research Laboratory) in Aberdeen.

Herman Goldstine was a mathematician at the University of Michigan before he enlisted in the US Army during the Second World War. He was about to be posted to the Pacific when Veblen, who was trying to recruit scientists for the BRL, stepped in with a better offer. Goldstine's orders to ship out arrived on the same day as those telling him to report to the Aberdeen Proving Ground. Goldstine sensibly chose Veblen and Aberdeen, where he was assigned to the team calculating artillery firing tables – essentially the same sort of projectile trajectories that Veblen had been hired to oversee during the First World War.

One evening in the summer of 1944, on the platform of Aberdeen train station, Goldstine saw someone he recognized – someone whose lectures he had once attended and who was by now the most famous scientist in America after Einstein. He introduced himself to von Neumann, and the two began to chat while waiting for their train. Goldstine explained that his role involved liaising with the Moore School in Philadelphia and mentioned the project they were working on together: an electronic computer capable of more than 300 multiplications per second.

At that, Goldstine says, 'the whole atmosphere of our conversation changed from one of relaxed good humour to one more like the oral examination for the doctor's degree in mathematics.'[8]

Quite soon after that meeting, on 7 August, according to Goldstine, he arranged for von Neumann to visit the machine being built at the Moore School. What he saw there, says Goldstine, 'changed his life for the rest of his days.'[9]

The ENIAC occupied a room that was roughly 30 feet wide and 56 feet long. Arrayed along the walls were the computer's innards: some 18,000 vacuum tubes, and banks of wiring and switches arranged in panels 8 feet tall.

'Now we think of a personal computer as one which you carry around with you,' says mathematician Harry Reed, who joined the project in 1950. 'The ENIAC was actually one that you kind of lived inside.'[10]

The ENIAC was the brainchild of John W. Mauchly, a former physics teacher whose dreams of a research career had been dashed

The ENIAC. *Courtesy of the University of Pennsylvania archives.*

by the Great Depression. He had won a scholarship to study at Johns Hopkins University in Baltimore, Maryland and without bothering to finish his undergraduate degree he earned a PhD in physics from there in 1932. He worked briefly as a research assistant but had the misfortune of beginning his hunt for a university post during one of the longest periods of economic decline in recent history. Mauchly's academic job hunt stalled. Instead, he had to settle for employment at Ursinus, a small liberal arts college in Pennsylvania, where he became the head and, indeed, the sole member of the physics department. He was still there when war broke out.

While one global catastrophe had thwarted Mauchly's ambitions, another now transformed his prospects for the better. In 1941, he took an electronics course at the Moore School, where scientists were being retrained to help with the war effort. There the thirty-four-year-old Mauchly met a twenty-two-year-old J. Presper Eckert, the electronics whizz-kid son of a local real estate magnate who was running the course teaching laboratory. Together, the pair cooked up ambitious plans for a machine to compute artillery firing tables, a task that was consuming an increasing proportion of the Moore School's resources.

A firing table for a particular gun consisted of hundreds of trajectories showing the ranges of shells fired at various elevations under different conditions. Each combination of gun and ammunition required its own firing table. The Moore School's calculations began with data from BRL's test range. About ten rounds were discharged at different altitudes, and the distance travelled by each shell measured and recorded. The Moore School then had to compute other trajectories, taking into account wind resistance, which varies with the height and speed of the shell. Calculating a single line in a firing table could take a person working with a desk calculator up to two days. By the 1930s, the same task could be completed in less than twenty minutes using a differential analyser, an ingenious tool invented by Vannevar Bush and his colleagues at the Massachusetts Institute of Technology (MIT). The analyser is a room-filling device that looks somewhat like a set of over-designed table football tables bolted together. Once set up with the appropriate shafts, gears and wheels for a problem, one arm of the analyser would be used to trace an input curve and the movements would be mechanically transformed into the required output. They did not come cheap. The Army had paid for the Moore School's analyser on condition that the BRL could requisition it if war threatened to break out. In 1940, it did and the BRL sent Lieutenant Goldstine to act as a liaison.

By the end of 1942, the Moore School had one group operating the analyser and a second team of a hundred women working with desk calculators six days a week to calculate trajectories. Both teams were each completing a firing table in about a month. But despite the women's efforts, they were falling further and further behind schedule.

Mauchly knew about the backlog. The team included his first wife, the mathematician Mary Augusta Walzl. After the Moore School hired Mauchly as an assistant professor in September 1941, he began sniffing around the analyser to learn what made it tick and started thinking about an electronic replacement that would do the job faster. He set out his preliminary ideas in a memo, *The Use of High Speed Vacuum Tubes for Calculating*. Some time in spring 1943, Goldstine came across the memo. He became convinced that the approach Mauchly described was worth pursuing and persuaded BRL officials of its importance. The contract for the ENIAC was signed in June with $150,000 set aside by the BRL for completing the project in fifteen months. The final cost of the computer was well over half a million – $8 million today.

Work on the ENIAC, now code-named 'Project PX', began in earnest. Electrical engineer John Brainerd was appointed the principal investigator of the project, charged with overseeing the budget, and Eckert became the project's chief engineer. Mauchly, who had conceived the project in the first place, was relegated to a part-time role as consultant. With so many of its faculty engaged in war work, the Moore School needed him to continue teaching.

Eckert initially led a small team of a dozen or so engineers who designed and tested the machine's circuits. But when construction began in 1944, the workforce quickly ballooned. A production team of thirty-four 'wiremen', assemblers and technicians were brought on board to install the machine's components, wire them together and solder the half million or so joints required to bring the computer into being. And though the ENIAC had been designed by men, this gruelling, fiddly job of actually building it was almost exclusively the work of women, who laboured nights and weekends until it was complete.[11] Buried in the project's payroll records are the names of nearly fifty women and perhaps many more who were only listed by their initials.

Despite their efforts, and to the barely contained frustration of Army chiefs, difficulties procuring parts during wartime – not just vacuum tubes but more mundane components like resistors, switches, sockets and miles of wiring – kept delaying the project's completion. 'There was about eighteen months where ENIAC was consistently three months from being finished,' says historian Thomas Haigh.[12]

By the time von Neumann arrived on the scene in August 1944, the ENIAC was still more than a year away from completion. One of his first contributions to the project was to keep the money flowing in. A scientific heavyweight, von Neumann was also by now a man of considerable influence in government and military circles. He argued convincingly that the machine's usefulness would extend far beyond the purpose for which it had been designed. When the machine was finally ready in December 1945, his prophecy came true. The ENIAC's first job was not firing table calculations at all but hydrogen bomb problems from Los Alamos.

Los Alamos dispatched two physicists, Nicholas Metropolis and Stan Frankel, to find out how the new machine worked so they could squeeze from it every last drop of computational power it could muster. They were aided by mathematician Adele Goldstine, Herman's wife, who would later write the ENIAC's user manual, and six newly trained operators – all women, four with mathematics degrees. Only the two physicists knew the real purpose of the calculations: to determine how much precious tritium would be required to ignite Teller's Super by solving a set of three partial differential equations.[13] Like much of America's early bomb research, the details remain classified, but according to Herman Goldstine, a million punched cards were shipped to the Moore School over the next several weeks. Teller used the results to press the case for his bomb: in a secret meeting held at Los Alamos in April 1946 he declared that the computer had vindicated him; his Super would work. It was the very same meeting that led von Neumann and Fuchs to collaborate on their patent, the details of which Fuchs would leak to the Russians.

*

Von Neumann's interest in the ENIAC went far beyond its usefulness as a tool for creating better bombs. From the first moment he saw it, he was thinking of a radically different kind of computer altogether.

Many of the ENIAC's drawbacks were recognized by its designers early on in the project. Of the 150 kilowatts of power it consumed, more than half was used to either heat or cool the tubes. Despite rigorous procedures for stress-testing new batches and rejecting those that did not pass muster, a tube blew every couple of days. To minimize the disruption caused by malfunctioning parts and broken connections, the ENIAC's components were arranged in standardized plug-in units that could quickly be removed and replaced. Even so, the machine was down much more often than it was up. In December 1947, by which time the ENIAC had been moved to the BRL in line with contractual obligations, a piece in the *New York Times* noted that 17 per cent of the time was spent on set-up and testing, 41 per cent on trouble-shooting and fixing problems and only 5 per cent – about two hours per week – on doing actual work.[14]

The ENIAC was born as a machine of war, made for a single purpose. But with the war over and firing tables competing for time on the machine with other pressing problems, the machine's *raison d'être* had become its greatest handicap. Von Neumann saw this more clearly than anyone on the project, perhaps more clearly than anyone in the world. More importantly, he understood exactly how to design a greatly more flexible successor to the ENIAC that could be easily reprogrammed. The ENIAC team had been discussing the shortcomings of their machine for some time. Now, with von Neumann on board, a proposal to build a successor was quickly prepared for review by the BRL's top brass. On 29 August, with Goldstine and von Neumann present, a high-level committee gave the plans their blessing. At the Moore School, work on the new machine was quickly codenamed 'Project PY', and intense discussions began over its design. By March the following year, the team agreed that von Neumann would summarize their thinking. He would do much more than that.

Von Neumann was by now well versed in the nascent field of electronic engineering. He had opinions on the relative merits of different vacuum tubes and was keen to design circuits for the new machine. But he was not an engineer – he was a mathematician, with an extraordinary capacity to cleave problems of their superficial complexities and render them in their most elemental form. He now brought this talent to bear on the jumbled ideas of the ENIAC team. 'John von Neumann was no aesthete,' note Haigh and his colleagues, 'but his intellectual response to ENIAC might be likened to that of a Calvinist zealot who, having taken charge of a gaudy cathedral, goes to work whitewashing frescos and lopping off ornamental flourishes.'[15] That impulse would yield a design for the new machine that would inspire generations of engineers and scientists to build computers in its image.

Curiously, von Neumann was mentally prepared for this cutting-edge contribution to computing by his involvement in the foundational crisis that had riven mathematics in the early twentieth century. An unlikely turn of history would entangle the intellectual roots of the modern computer with Hilbert's challenge to prove that mathematics was complete, consistent and decidable. Soon after Hilbert issued his challenge, the intellectually dynamic but psychologically frail Austrian mathematician Kurt Gödel would demonstrate that it is impossible to prove that mathematics is either complete or consistent. Five years after Gödel's breakthrough, a twenty-three-year-old Turing would attack Hilbert's 'decision problem' (*Entscheidungsproblem*) in a way completely unanticipated by any other logician, conjuring up an imaginary machine to show that mathematics is not decidable. The formalisms of these two logicians would help von Neumann crystallize the structure of the modern computer. The result of his musings, *First Draft of a Report on the EDVAC*, would become the most influential document in the history of computing.[16] 'Today,' says computer scientist Wolfgang Coy, 'it is considered the birth certificate of modern computers.'[17]

Just over a decade after Hilbert's quest for a perfectible mathematics had run aground, his programme would unexpectedly bear spectacular fruit.

The story of Gödel's surprising contribution to computing began in 1930 at the tail-end of a three-day conference on the foundations of mathematics held in Königsberg and attended by many of the field's worthies, including the young von Neumann.[18] Though the Austrian logician's work would ultimately destroy Hilbert's programme, he had set out to advance it. For his doctoral dissertation,

Gödel had pursued one of the basic proofs demanded by Hilbert – to show that 'first-order logic' (also known as the 'predicate calculus') was complete.

First-order logic is a set of rules and symbols that can, for example, express the sort of formal arguments or 'syllogisms' found in the classical logic of Aristotle.[19] The beauty of this sort of symbolic logic is that the system formalizes natural language, stripping the statements down to their bare logical bones. Hilbert had helped to develop 'first-order logic' in a series of lectures between 1917 and 1922. With his student Wilhelm Ackermann he produced a classic textbook on the subject in 1928.[20] What Hilbert knew was lacking, and requested in his book, was a rigorous proof that theorems in this language were reliable. Of course it is possible to say some pretty odd things with first-order logic.[21] But if one started from premises that were true, Hilbert asked, are the conclusions reached by first-order logic always true as well? Or, in mathematical terminology, is first-order logic 'complete'? This is exactly what the twenty-three-year-old Gödel had set out to prove in his thesis.

On the second day of the Königsberg conference, Gödel gave a twenty-minute talk that outlined his proof, showing that first-order logic was, indeed, complete. Yet impressive as the work was for a young logician cutting his teeth on the subject, the result was that expected by the community and the methods he used were well established. Gödel's presentation met with nodding approval from the assembled luminaries of mathematics and philosophy. But the young mathematician was hiding an ace up his sleeve.

Only on the last day of the conference, towards the end of a round-table discussion that would bring the meeting to a close, did Gödel play it. In a single sentence, he quietly reduced the earlier proceedings of the conference to an irrelevance and the foundations of mathematics to a flaming ruin. 'One can,' he ventured modestly, 'even give examples of propositions (and in fact of those of the type of Goldbach or Fermat) that, while contentually true, are unprovable in the formal system of classical mathematics.'

In other words, there are truths *in* mathematics that cannot be proven *by* mathematics. Mathematics is not complete. His off-hand references to Goldbach and Fermat were portentous. Golbach's conjecture (that all even numbers greater than 2 are the sum of two primes) and Fermat's last theorem (that no positive integers a, b, and c satisfy the equation $a^n + b^n = c^n$ for values of n greater than 2) were two of the great unsolved problems in arithmetic.[22] Gödel was implying that even in the sort of maths taught to schoolchildren there might lurk verities that can never be substantiated.

Gödel had announced one of the foremost intellectual feats of the twentieth century. He must have been rather piqued when the response to this revelation was even more muted than the perfunctory praise that had greeted his thesis. The assemblage politely ignored his carefully prepared one-sentence bombshell as if he had cracked a bad joke at a dinner party.

Gödel's proclamation ought at least to have prompted a question or two. How could, for instance, a mathematical proposition be true yet unprovable? Yet it appears that only one person grasped the import of Gödel's achievement well enough to want to know more. After the round table was over, von Neumann, there as a trusted evangelist for Hilbert's programme, grabbed Gödel by the sleeve and steered him to a quiet corner to grill him about his methods.

The following day Hilbert gave his retirement address, a passionate speech in which he declared again there were no unsolvable problems in mathematics and uttered the words that would become the epitaph on his gravestone: *We must know – we will know*. But Gödel had already proved him wrong.

At the heart of Gödel's proof was a restatement of the liar's paradox. In its usual form, the paradox can be phrased: 'This sentence is false.' Awkwardly, this is only true if it is false – a fact that has tied grammarians and logicians in knots for centuries. Gödel recast this in terms of unprovability rather than falsity, devising an analogous statement – call it 'G' – that says: 'the proposition g is not provable in the system'.

This is not in itself paradoxical. But then Gödel turned the statement back on itself by making the subject, g, of the proposition be G itself. Now, if the proposition G cannot be proved, then it is true. If, on the other hand, G *can* be proved, then it is false. But that would mean the truth of the original statement – that G is not provable – had been formally demonstrated. Thus, G is also true. A mathematical system in which the same statements can be both true and false would be useless. So better to plump for the first option: there are true but unprovable statements in mathematics. One implication of this that Gödel himself would champion in later life was the Platonic idea that mathematical truths were in some sense 'out there' – they are discovered, not invented.

Gödel had done much more than conjure up yet another logical paradox. His entire proof was written in the language of arithmetic – it was a mathematical theorem. He created an ingenious system, now called Gödel numbering, of assigning numbers to logical statements then manipulated them in strict accordance with the rules and axioms of arithmetic as set out in the *Principia*.

Anyone can come up with an arbitrary system capable of converting the letters and symbols of formal logic to numbers. Perhaps as a child, you wrote messages in a 'secret' code for your friends, with each letter of the alphabet assigned a number. Gödel began the same way, assigning a unique number to each symbol of the *Principia*.[23] This is where the resemblance to any primitive coding antics ends because Gödel next used a set of arithmetic rules to assign every statement, and every collection of statements (including those that form theorems), a unique number. The process is reversible so that every Gödel number can be decoded back into the original expression.

For his proof, Gödel ensured every logical operation on a statement entailed a parallel arithmetic operation on its number. So, imagine a statement, 'all oranges are green', has a Gödel number of 3778 (actual Gödel numbers are far longer). The negation of that statement, 'not all oranges are green' might have a Gödel number double that: 7556. If this were really Gödel's system, we would find on decoding the number 7556 that it did really represent the statement 'not all oranges are green'. Furthermore, doubling the Gödel number of *any* statement would *always* yield its negation in this way. Incredibly, Gödel ensured *all* logical operations in his system had an arithmetic equivalent. So the Gödel numbers of statements that follow one another (a syllogism, for example) are linked by arithmetic relationships, just as the statements themselves are linked by the rules of logic. A proof comprises a set of logical inferences of this sort. At the root of a proof are a few axioms – themselves logical statements.

Any valid proof must have a Gödel number which can be derived from the Gödel numbers of the axioms using the rules of arithmetic.

With his coding system established, Gödel was able to provide a process whereby any proof could be checked by simple maths. First, decompose the Gödel number of the proof by following the decryption rules of Gödel's scheme. Eventually this will reveal the Gödel numbers of the proof's axioms. Then simply check that the Gödel numbers of the axioms are indeed those that are allowed by the system (i.e. the axioms of the *Principia*). The process is repetitive, so Gödel defined a combination of 'primitive recursive functions' – essentially, mathematical loops – to do the decoding, producing an algorithm, a sort of proof-checking machine, that could test any theorem. The question of whether a theorem is valid becomes a question of doing sums (albeit quite complicated ones!). Put the Gödel number of the theorem into the machine, turn the handle until the Gödel numbers of the axioms pop out, then consult the list of the system's foundational axioms to see if they are there.

Finally, Gödel produced an arithmetic statement – a complicated sum if you like – that mirrored the phrase 'the proposition *g* is not provable in the system'. He then demonstrated that the Gödel number of this phrase could *itself* be *g*. That is, Gödel had added another layer of self-reference to the original statement by making arithmetic talk about the very nature of arithmetic. His breathtaking discovery was that the language *of* mathematics could be used to make meta-statements *about* mathematics.

Generations of philosophers and mystics have made much of Gödel's theorem, each claim wilder than the next. The cognitive scientist Douglas Hofstadter has seen in that loop of self-reference a glimmer of our self-awareness, the very essence of human consciousness.[24] Some have even suggested that the work provides evidence for the existence of God (who had left these truths floating free for mathematicians to discover?), claims bolstered by the discovery of an unfinished proof for the existence of God in Gödel's papers after his death.

But one remarkable consequence of his arcane paper is not widely appreciated. In 1930, Gödel had written a computer program long before any machine capable of running it would exist.[25] He had dissolved in one fell swoop the rigid distinction between syntax and data. He had shown that it was possible to devise a rigorous system in which logical statements (that were very much like computer commands) could be rendered as numbers. Or as von Neumann would put it in 1945 while describing the computer he was planning to build at the IAS, '"Words" coding the orders are handled in the memory just like numbers.'[26] That is the essence of modern-day coding, the concept at the heart of software. Any program is converted to binary machine code before it is executed line by line by a computer's central processing unit, each number corresponding to a task that is hardwired into the chip. The primitive recursive functions in Gödel's proof appear in modern computer programs as 'for-loops', used to repeatedly execute a block of code. The numerical addresses used to reference items in computer memory are reminiscent of Gödel numbers, which, among other things, act as tracking codes for logical statements.

'Someone knowledgeable about modern programming languages today looking at Gödel's paper on undecidability written that year will see a sequence of forty-five numbered formulas that looks very much like a computer program,' the mathematician Martin Davis explains. 'The resemblance is no accident. In demonstrating that the property of being the code of a proof in PM [*Principia Mathematica*] is expressible *inside* PM, Gödel had to deal with many of the same issues that those designing programming languages and those writing programs in those languages would be facing.'[27]

Von Neumann kept thinking about Gödel's proof after the Königsberg conference. On 20 November, he wrote excitedly to Gödel. 'Using the methods you employed so successfully . . . I achieved a result that seems to me to be remarkable, namely,' von Neumann continued with a flourish, 'I was able to show that the consistency of mathematics is unprovable.'

Von Neumann promised to send him his proof, which he said would soon be ready for publication. But it was too late. Gödel, probably sensing that von Neumann was hot on his heels after their conversation in Königsberg, had already sent his paper to a journal.[28] He now sent a copy to von Neumann. Crestfallen, von Neumann wrote back, thanking him. 'As you have established the theorem on the unprovability of consistency as a natural continuation and deepening of your earlier results,' he added, 'I clearly won't publish on this subject.' So saying, von Neumann quietly passed up the opportunity to stake a claim on the most remarkable result in mathematical history.

The consequences of Gödel's head-spinning second incompleteness theorem were even more staggering than his first. As von Neumann had indicated, building on Gödel's earlier theorem, no system complex enough to contain arithmetic could be proven to be consistent – at least, not using the tools of the system itself. That is, Gödel had shown that it is impossible to prove that statements that contradict our common-sense notions of the counting numbers (like 2+2 = 5) can themselves never be proved.

While others were still grappling with Gödel's work, von Neumann immediately understood what Hilbert, and even Gödel himself, would struggle to accept. 'My personal opinion,' he said, 'which is shared by many others, is that Gödel has shown that Hilbert's program is essentially hopeless.' Von Neumann called Gödel the greatest

Kurt Gödel.

logician since Aristotle and gave up working on the foundations of mathematics. When, after Austria was annexed into Nazi Germany in 1938, Gödel was denied a post at the University of Vienna partly for having 'always travelled in liberal-Jewish circles', von Neumann would campaign to bring him to the IAS. 'Gödel is absolutely irreplaceable; he is the only mathematician alive about whom I would dare to make this statement,' he wrote to Flexner. 'Salvaging him from the wreck of Europe is one of the great single contributions anyone could make to science at this moment.'[29]

He was successful. In Princeton, Gödel became convinced that the radiators and refrigerator in his home were emitting poisonous gas and had them removed, shivering through his first few winters. He would sometimes wander into von Neumann's house in his absence, pick up a book and start reading. Once he had finished, he would replace the book and leave – all without saying a word to von Neumann's young wife. He often walked home from the institute with Einstein, engrossed in gentle disagreements about politics, physics and philosophy. The solemn young logician and the gregarious physicist, now in his sixties, made an odd couple but nonetheless derived great pleasure from each other's company. After the deaths of his friends Einstein and von Neumann in the 1950s, Gödel largely ceased to publish original work. In 1978, with his wife, Adele, seriously ill in hospital, Gödel suffered one of the many severe bouts of paranoia that had afflicted him through life. Gripped by a fear of being poisoned, he refused to eat and slowly began to starve. When Adele recovered and returned home five months later, she quickly convinced the near-skeletal Gödel to enter a hospital. It was too late. Gödel died on 14 January 1979, weighing barely 66 pounds.

In 1931, Gödel demonstrated that mathematics could not be proven to be either complete or consistent. Five years later, Turing produced his negative answer to the last of Hilbert's three questions – is mathematics decidable?[30] Just as he was preparing his proof for publication, a paper by the American logician Alonzo Church arrived in Cambridge. Church had used his own formal system of logic, called 'lambda calculus', to arrive at his, equivalent, answer to the *Entscheidungsproblem*.[31]

Under normal circumstances, Church's paper would have rendered the younger logician's work more or less unpublishable. There are rarely prizes for second place in mathematics. But Turing's approach to the problem was so startlingly novel that Max Newman, his Cambridge mentor, recommended he press ahead with publication – and asked Church to review the paper. Church assented and even agreed to supervise Turing's PhD from the following year in Princeton.

In his famous paper, Turing described an imaginary machine that can read, write or erase symbols on an infinitely long tape, divided into squares.[32] The machine's head can move left or right along the tape, one square at a time, reading the symbol it finds there. Depending on what it finds, the machine executes one or more actions from its limited repertoire of print, erase and move. What it does at each square is determined by what Turing called the machine's '*m*-configuration', its internal state, which can also change in response to the contents of the square being read. Turing presents a simple example of such a machine which prints the binary sequence 01010101 . . . forever to an empty tape, starting at any blank square and leaving a blank square between adjacent digits.[33]

Given the appropriate instructions, Turing's imaginary machines can add, multiply and carry out other basic mathematical operations – though Turing does not bother to demonstrate this.[34] He instead builds sets of instruction tables for a variety of ancillary tasks such as searching for and replacing a symbol or erasing all symbols of a particular type. Later in the paper, he uses these to help build a 'universal computing machine' that is capable of simulating any other Turing machine. Computer programmers today would recognize Turing's strategy: modern programmes make use of libraries of simpler programmes known as 'subroutines'. Subroutines simplify the structure of programs, and simpler programs are easier to understand, improve and troubleshoot.

Though Turing describes his computing machine in purely abstract terms, it is quite easy to imagine building one. A Turing machine that can execute a single task might comprise a scanner, print head (erasing characters is admittedly a little trickier) and a motor that moves a 'limitless' roll of tape backwards and forwards. The *m*-configurations, along with their possible inputs and outputs, are 'hard-wired' into the device, so when the scanner reads a symbol, the print head moves, erases or prints in accordance with the instruction table.

Such machines are called 'program-controlled' computers. A modern washing machine is a program-controlled computer. So was the ENIAC (though, unlike a washing machine, it could be reconfigured to tackle other jobs by plugging patch cables into appropriate sockets like an old-fashioned telephone switchboard). In both cases, the push of a button or flick of a switch sets off a chain of computational events that transforms one thing (dirty laundry, punched cards bearing firing range data) into another (clean laundry, punched cards with calculated artillery trajectories).

The universal machine Turing described, however, is quite different. When fed the instruction table of another Turing machine, the universal machine can simulate it exactly. Turing first describes a system to convert an instruction table into a form that his machines can digest: a stream of letters that can be written to tape. Turing calls this a 'standard description' of the machine. Today, we might call it a 'program' – the type that is stored in a computer's memory rather than wired in.[35]

Painstakingly 'built' from the library of subroutines and instructions that Turing had defined, the universal machine is complicated but of finite size. Turing is able to provide a complete description of this remarkable device in a little over four pages. Yet given the appropriate instructions, the universal machine can execute an infinite variety of tasks.

'We might expect a priori that this is impossible. How can there be an automaton which is at least as effective as any conceivable automaton, including, for example, one of twice its size and complexity?' von Neumann asked in 1948. 'Turing, nevertheless, proved that this is possible.'[36]

So celebrated is Turing's invention of the universal machine, the reason he invented it – to solve Hilbert's *Entscheidungsproblem* – is often overlooked. Nonetheless, the whole logical apparatus of Turing's paper was assembled for the sole purpose of answering it.[37] With it, he proves there can be no general, systematic process for deciding whether or not any particular statement of first-order logic is provable, dashing the last of Hilbert's dreams.

When Turing wrote his paper, 'computers' were not machines but human beings. In fact, they were nearly always women working with desk calculators, pen and paper (though Turing uses male pronouns in referring to them). Turing had modelled his machines on these human computers (a machine's *m*-configuration, for instance, corresponds to a 'state of mind'). In the last part of *Computable Numbers* Turing argues that his machines are capable of carrying out any algorithmic process that can be performed by their human counterparts. Conversely, a human may compute anything a Turing machine can – assuming the human in question does not die of boredom first – but nothing that the machine cannot.

The universal Turing machine is now considered an abstract prototype of a general-purpose 'stored program' computer – one that can, like any laptop or smart phone today, execute an application in the computer's memory. The older machines 'could only play one tune . . . like a music box,' says Klári. 'In contrast, the "all purpose" machine is like a musical instrument.'[38]

From the 1950s onwards, the universal machine has been appropriated as a foundation stone of theoretical computer science. So often is Turing's name mentioned in the same breath as the programmable computer, the myth has sprung up that he somehow *invented* the computer. The difficulties of distinguishing myth from reality are exacerbated by the fact that Turing *did* design a computer (the ACE, for the National Physical Laboratory in 1945, after he saw von Neumann's EDVAC report), and by Turing's very real formative contributions to the field of artificial intelligence. But 'On Computable Numbers' was a work of abstract logic; its avowed purpose to solve Hilbert's decision problem. Like Schrödinger's famous description of an alive-dead cat, published exactly twelve months earlier, Turing's machine was a thought experiment. 'Schrödinger was not trying to advance the state of the art of feline euthanasia,' says Haigh. 'Neither was Turing proposing the construction of a new kind of calculating machine.'[39]

Indeed, perhaps Turing's only contribution to the practice of contemporary computing was through his influence on an older colleague at Fine Hall. Von Neumann would sing the praises of Turing's work, later pressing the engineers working on his computer project at the IAS to read the Englishman's obscure paper. 'Von Neumann brought his knowledge of "On Computable Numbers" to the practical arena of the Moore School,' says Copeland, perhaps Turing's most passionate advocate. 'Thanks to Turing's abstract logical work, von Neumann knew that by making use of coded instructions stored in memory, a single machine of fixed structure could in principle carry out any task for which an instruction table can be written.'[40] With the EDVAC report of 30 June 1945, von Neumann would turn Gödel and Turing's abstract musings into the canonical blueprint for the stored-program computer.

Von Neumann's *First Draft of a Report on the EDVAC* is a curious document. He mentions electronic components mostly to explain why he will not be discussing them: his aim is to describe a computer system without getting bogged down in the specifics of engineering. 'In order to avoid this we will base our considerations on a hypothetical element, which functions essentially like a vacuum tube,' he says.[41] His 'hypothetical element' is an idealized neuron, shorn of its physiological complexities. This seems odd today, but von Neumann, Turing, Norbert Wiener and other thinkers who contributed to the foundations of the field that became known as 'artificial intelligence' *did* think about computers as 'electronic brains'. Today using 'brain' or 'neuron' in the context of computers seems laughably naive. Yet we accept the similarly anthropomorphic use of 'memory' to mean 'storage' without blinking an eye.

The idealized neuron of von Neumann's EDVAC report came from work published by the neurophysiologist Warren McCulloch and the mathematician Walter Pitts in 1943.[42] What they described was a vastly simplified electronic version of a neuron, which summed a number of input signals together and fired off a signal if that sum exceeded a certain threshold. A real neuron is a lot more complicated than this, for instance summing thousands of input signals and producing a train of pulses rather than a single blip. McCulloch and Pitts argued that neurons could nonetheless usefully be treated as switches. They showed that networks of such model neurons could learn, calculate, store data and execute logical functions – they could, in short, compute. Whether they had 'proved, in substance, the equivalence of all general Turing machines – man-made or begotten', as McCulloch later claimed, is a point of contention even today.

Von Neumann adopted the terminology and notation of McCulloch and Pitts to describe for the first time the structure of a stored-program computer. Theirs was the only paper he would cite in the whole of the EDVAC report. There were five distinct parts or 'organs' in the assembly that von Neumann described. The first three components were a 'central arithmetic' unit for performing mathematical operations such as addition and multiplication; a 'central control' unit to ensure that instructions were executed in the proper order; and a 'memory', a single organ that would store both computer code and numbers. The fourth and fifth components were the input and output units, for shipping data in or out of the machine. In keeping with the spirit of the McCulloch and Pitts paper, von Neumann drew parallels between his machine and the human nervous system. The input and output units respectively corresponded to sensory neurons, bringing signals from receptors to the central nervous system, and motor neurons, carrying nerve impulses back to muscles and organs. In between were associative neurons, which modulated the signal – an equivalent task to that discharged by the first three units he had specified in his report (central arithmetic unit, central control and memory).

When Goldstine received the report, he was in raptures. He congratulated von Neumann for providing the first 'complete logical framework for the machine' and contrasted the streamlined design with the ENIAC, which was 'chuck full of gadgets that have as their only raison d'etre that they appealed to John Mauchly'.[43]

Computer designers now refer to the whole configuration as the 'von Neumann architecture', and nearly all computers in use today – smart phones, laptops, desktops – are built according to its precepts. The design's fundamental drawback, now called the 'von Neumann bottleneck', is that instructions or data have to be found and fetched serially from memory – like standing in a line, and being able to pass messages only forwards or backwards. That task takes much longer than any subsequent processing. That handicap is outweighed by the architecture's considerable advantages, which stem from its simplicity. The ENIAC had twenty modules that could add and subtract, for example. In the EDVAC, there would be one. Less circuitry means less that can go wrong, and a more reliable machine.

Von Neumann emphasized that the memory of the machine should be large, citing the sort of number-crunching capabilities that he knew would be required by Los Alamos. The EDVAC report called for the computer to have a storage capacity of 8,000 words, each 32 bits long – much more than the ENIAC, which could store only twenty ten-digit numbers. 'Just as ENIAC was shaped by the firing table problem,' historians note, 'EDVAC was shaped by the fluid dynamics of the atom bomb.'[44]

A large section of the report was devoted to the delay line, a cheap high-capacity electronic memory that Eckert had invented early in 1944. Before meeting Mauchly, he had applied his electrical engineering wizardry to the wartime development of radar systems. One goal of that work had been to find reliable methods of distinguishing between signals from stationary and moving objects. Eckert's ingenious solution was to turn the incoming electrical radar signal into sound waves and pass them through a tube filled with mercury. The signal could be recovered at the other end of the tube by converting the sound waves back into electrical impulses. The length of the tube was chosen so that the signal was delayed long enough for the radar dish to turn through one complete revolution. The delayed signal could then be subtracted from the new signal to leave only blips from objects in motion. While working on the ENIAC, Eckert realized these 'mercury delay-lines' could be used to store and retrieve data. Data could be stored more or less indefinitely in the tubes by feeding exiting signals back into the device and could be overwritten by 'catching' the relevant set of pulses as they emerge from the tube and replacing them with the new data. Depending on its length, a delay line could store hundreds of digits.

Around 11,000 of the ENIAC's 18,000 temperamental vacuum tubes were given over to storage. The first generation of EDVAC-style computers would need around ten times fewer because they stored digits by circulating acoustic signals in tubes of mercury. The delay line would quickly be superseded. First, cathode ray tubes would store digits as dots of charge on a phosphor screen. Then magnetic-core memory would flip the magnetization of ceramic rings to store either a one or a zero. This in turn gave way to semiconductor memory chips with banks of tiny transistors and capacitors that together hold millions of times more information than those early tubes of mercury. But the working memory of computers today is based on the same principle as Eckert's delay lines: any device that can hold data temporarily can hold it in perpetuity if the bits stored on it are refreshed regularly.[45]

There would never be a finished version of the EDVAC report. In the summer of 1945, von Neumann had more pressing commitments. The 'first draft' he sent to Goldstine stopped abruptly and contained blank spaces that had been left to add references and notes. Descriptions of the input and output units were sparse, and von Neumann later sent details of how programs might be recorded and read from magnetic tapes, to be included in a future second draft. Goldstine, however, was delighted with the incomplete report and quickly had it typed up. Without informing von Neumann, Mauchly or Eckert, he sent the draft to dozens of scientists and engineers who were busily planning their own computers in America and elsewhere. Among those inspired was Turing, who would cite the report nine months later in his own plans for a computer, the Automatic Computing Engine (ACE).

Not everyone was pleased. The report Goldstine circulated had only the name of John von Neumann on its title page. Eckert and Mauchly, who were hoping to patent aspects of computer design, were furious. The ENIAC's inventors accused von Neumann of inflating his contribution to the project and rehashing their work. 'Johnny learned instantly, of course, as was his nature,' Mauchly wrote. 'But he chose to refer to the modules we had described as 'organs' and to substitute hypothetical 'neurons' for hypothetical vacuum tubes or other devices which could perform logical functions. It was clear that Johnny was rephrasing our logic, but it was still the SAME logic. Johnny did NOT alter the fundamental concepts which we had already formulated for the EDVAC.'[46]

The pair blamed von Neumann for cheating them out of millions of dollars, and their opprobrium grew when they learned he had made thousands himself by consulting for one of their greatest commercial rivals. Their bitterness lived on long after von Neumann's death. 'He sold all our ideas through the back door to IBM,' Eckert complained in 1977.[47]

Arthur Burks, a mathematician turned engineer who worked on

both the ENIAC and von Neumann's IAS computer, later scrutinized the written records from the period to trace the evolution of the stored program idea. 'I do not think any of us at the Moore School had an architectural model in mind for the EDVAC until we learned of Johnny's,' he concluded.[48] Goldstine, who would also work with von Neumann at Princeton, agreed. 'It is obvious that von Neumann, by writing his report, crystallized thinking in the field of computers as no other person ever did. He was, among all members of the group at the Moore School, *the* indispensable one.'

Eckert and Mauchly's accusations were not, however, without some justification. It is probably true, for instance, that von Neumann deliberately did not mention the ENIAC project anywhere in the EDVAC report. The ENIAC was still a state secret, and its omission allowed his draft to be circulated widely. Later, though he never claimed to be the only brain behind EDVAC, von Neumann did not go out of his way to deny it. The money no doubt helped von Neumann salve his conscience, as Eckert and Mauchly implied. For several years, IBM paid von Neumann the equivalent of nearly a whole year's salary (which, of course, he continued to draw from the institute) for just thirty days of consulting. But von Neumann only began working for IBM in 1951. When he was first approached by the company in 1945, he was intent on securing funding for his own computer project and turned them down. Had money been his primary concern, why wait six years before accepting IBM's offer?

There is evidence that von Neumann had a nobler motive for playing down Eckert and Mauchly's roles. He wanted to accelerate the development of computers and feared the commercial route the ENIAC's inventors were pursuing would stifle progress with trade secrets and litigation. When von Neumann learned about the pair's claims in March 1946, he bristled, arguing that he was either joint or sole inventor of some of the EDVAC patents. 'I would never have undertaken my consulting work at the University had I realized that I was essentially giving consulting services to a commercial group,' he complained to his lawyer soon afterwards.[49]

The purpose of the EDVAC report, von Neumann testified the following year, was 'to contribute to clarifying and coordinating the thinking of the group' and 'further . . . the art of building high speed computers' by disseminating the work as quickly and as widely as possible. 'My personal opinion was at all times, and is now, that this was perfectly proper and in the best interests of the United States.'[50]

'I certainly intend to do my part to keep as much of this field 'in the public domain' (from the patent point of view) as I can,' von Neumann wrote to Frankel as he made plans for building his own computer at the IAS.[51] Patent rights to the IAS machine were in large part handed over to the government in mid-1947. The IAS team sent a stream of detailed progress reports to about 175 institutions in several different countries, helping to spawn a generation of computers across the world. 'The remarkable feature of the reports,' noted I. J. Good, who worked with Turing during and after the war, 'was that they gave lucid reasons for every design decision, a feature seldom repeated in later works.'[52]

The battle for ownership of the intellectual property and patent rights relating to the ENIAC and EDVAC would drag on for decades. Von Neumann would have been satisfied by the judge's eventual verdict, delivered on 19 October 1973. The automatic electronic digital computer was held to be in the public domain.

American law allows inventors up to a year to file after an invention is shown to be in working order. The ENIAC had started work in December 1945 with the Los Alamos calculations and been formally unveiled in February 1946, an occasion trumpeted by the front page of the *New York Times*. But patents were filed only on 26 June 1947, more than twelve months after the public disclosure of the invention. Furthermore, the judge held that von Neumann's EDVAC report, in circulation from mid-1945, had in any case divulged the ideas needed to build an electronic digital computer even earlier. In a last turn of the knife for Eckert and Mauchly, the judge deemed that they had not been the first to invent an automatic digital computer at all 'but instead derived that subject matter from one Dr. John Vincent Atanasoff'. The American physicist and his graduate student had built an electronic digital computer at Iowa State College. The Atanasoff–Berry computer (ABC) comprised 280 triodes and stored binary numbers using some 3,000 capacitors mounted on a pair of co-axial rotating drums. Weighing a mere 700 pounds, the ABC was a desk-sized minnow in comparison to the ENIAC and was neither a general-purpose nor a stored-program computer (its sole purpose was to solve sets of simultaneous equations). Mauchly always denied being influenced by it. Nonetheless, it is true that a few months before starting his electronics course at the Moore School in 1941, he had travelled to Ames to see Atanasoff's machine at first hand, familiarizing himself with its design and documentation.

What had become the longest trial in the history of the federal court system concluded with the ruling that the most valuable invention of the twentieth century could not be patented. The open source movement, born a decade or so later, would soon shun corporate secrecy, lauding the benefits of freely sharing information to drive forward innovation. Thanks to von Neumann those principles were baked into computing from the very beginning.

Von Neumann's regular visits to the Moore School ceased after 1946. By this time, the original ENIAC team had split, torn apart by the dispute over patent rights. The EDVAC is best known as a theoretical machine, described in von Neumann's report. Nevertheless a real computer called the EDVAC was shipped to the BRL in 1949. Without the leadership of Eckert and Mauchly or of von Neumann and Goldstine, the machine had become steadily more baroque, amassing more vacuum tubes as successive project teams moved away from the austere architecture described in the *First Draft*. Beset by technical problems, the bloated EDVAC only began computing anything useful three years after its arrival at Aberdeen. By then, it had been overtaken by other computers built according to von Neumann's radical design.[53]

Von Neumann was assailed by offers from prestigious universities wanting to host his new computer project. Norbert Wiener, hoping to lure his friend to MIT, asked how his plans would 'fit in with the Princetitute? You are going to run into a situation where you will need a lab at your fingertips,' he added, 'and labs don't grow in ivory towers.'[54]

Chicago offered him a professorship and a new institute to lead. Columbia and Harvard came knocking too. Realizing the IAS could be about to lose one of its brightest stars, Frank Aydelotte, the institute's far-sighted director, convinced the board of trustees to make $100,000 available to von Neumann immediately. 'Scholars have already expressed great interest in the possibilities of such an instrument,' von Neumann assured Aydelotte, 'and its construction would make possible solutions of which man at the present time can only dream.'[55]

The remainder of the cash, about $200,000, von Neumann secured principally from military sponsors, after convincing generals and admirals that advances in computing would be indispensable for activities such as designing jets or missiles, modelling the responses of an enemy during naval operations and, of course, building a superbomb. Remarkably, he persuaded them that all details of the project should be made public by arguing that the surest and fastest path to the powerful machines they needed would be through allowing others to learn from his embryonic project.

Von Neumann's priorities were now clear. Around this time, he met an old friend, Gleb Wataghin, a Russian-Italian physicist who had recently returned from Brazil. 'I suppose you are not interested in mathematics anymore,' Wataghin teased. 'I hear you are now thinking about nothing but bombs.'[56]

'That is quite wrong,' von Neumann replied. 'I am thinking about something much more important than bombs. I am thinking about computers.'

The 'Electronic Computer Project' did not, however, receive a warm welcome at the institute in Princeton, as Wiener had guessed. 'He clearly stunned, or even horrified, some of his mathematical colleagues of the most erudite abstraction, by openly professing his great interest in other mathematical tools than the blackboard and chalk or pencil and paper,' Klári says. 'His proposal to build an electronic computing machine under the sacred dome of the Institute, was not received with applause to say the least.'[57]

Their resistance was futile. Von Neumann pushed the project forward with the full force of his personality, touting the benefits of the machine to anyone who would listen:

> I think it is soberly true to say that the existence of such a computer would open up to mathematicians, physicists, and other scholars areas of knowledge in the same remarkable way that the two-hundred-inch

> telescope promises to bring under observation universes which are at present entirely outside the range of any instrument now existing.[58]

Whereas with the Army and Navy, he had stressed the near-term practical uses of computers, to his fellow scientists and to the IAS board he emphasized the discoveries that would be made – while remaining strategically vague concerning what they might be. 'The projected device, or rather the species of devices of which it is to be the first representative, is so radically new that many of its uses will become clear only after it has been put into operation,' he explained to one trustee. 'These uses which are not, or not easily, predictable now, are likely to be the most important ones. Indeed they are by definition those which we do not recognise at present because they are farthest removed from what is now feasible.'[59]

The IAS project began to gather momentum slowly. Von Neumann's first two hires were Goldstine, as director, and Burks. Space was limited at the IAS immediately after the war, in part because Aydelotte had granted refuge to League of Nations staff and their family during and immediately after the war. Goldstine and Burks were forced to squeeze in next to Gödel's room. The office they occupied was meant for Gödel's secretary, but the paranoid, withdrawn mathematician had never hired one.

Eckert could not be persuaded to join the IAS project, choosing instead to go into business with Mauchly. So von Neumann sought another chief engineer and found Julian Bigelow, who had been employed by MIT to make some of Wiener's impractical wartime ideas more practicable. Bigelow's most visible credential for the job was the car he drove to the interview at von Neumann's home – an ancient jalopy that only an engineer of the highest calibre could have made remotely roadworthy. The growing team now had to be housed somewhere. Aydelotte gave them the boiler room in the basement, conveniently located next to a men's lavatory. Funds for a new building to house the computer were secured and by the Christmas of 1946, the Electronic Computer Project was accommodated in a squat one-storey structure hidden behind a curtain of trees in a field some distance away from the institute's other buildings.

The team had a roof over their heads, but the IAS machine was not

The IAS computer project team, 1952. *Alan Richards photographer. From the Shelby White and Leon Levy Archives Center, Institute for Advanced Study, Princeton.*

materializing as fast as von Neumann had hoped. He had initially thought that a team of ten people could get the job done within three years. In reality, his computer would not be working until 1951. Los Alamos pressed von Neumann to find computational resources for a huge backlog of nuclear bomb calculations. In April 1947, recognizing that his machine would not be completed for some years yet, von Neumann hit upon the idea of converting the ENIAC into a primitive stored-program computer of the type he had described in the EDVAC report.

Von Neumann, together with Herman and Adele Goldstine, planned the conversion of the ENIAC.[60] Though she was only twenty-six years old, Adele held a Master's degree in mathematics and knew the ENIAC inside out. By July she had prepared a conversion plan that provided users with fifty-one different commands for writing programs and the detailed wiring changes and switch settings that would allow the ENIAC to decode and execute them. A team led by Jean Bartik (born Betty Jean Jennings), one of the ENIAC's original operators, was hired in March 1947 to programme the machine in its new guise – the first time anyone had been employed solely for that task. The job of computer programmer was born. The work from Los Alamos, however, needed people who could be trusted with nuclear secrets and understood the ENIAC's foibles. One of the few people who fitted that description was von Neumann's wife.

Klári describes herself as a 'mathematical moron' in her memoirs, but her view of herself was coloured by profound insecurity, no doubt exacerbated by a marriage to quite probably the smartest man on Earth. Klári, says Ulam, was 'a very intelligent, very nervous woman who had a deep complex that people paid attention to her only because she was the wife of the great von Neumann, which was not true of course'.[61]

With no education beyond high school (she had been sent to an English boarding school as a teenager), she took on a wartime job under Frank W. Notestein, the founding director of Princeton University's Office of Population Research. Notestein's group would become famous for their demographic projections, looking at the shift in the postwar populations of Europe and Russia, for example. This was hardly a job for someone as hopelessly innumerate as Klári claimed to be – especially as she appears to have thrived there: she was quickly promoted and turned down the offer in 1944 of an academic post at the university. Her interests were soon to turn from predicting the movements of people to predicting the movements of neutrons inside a hydrogen bomb.

'Bring riding and skating things if possible opportunities very good,' cabled von Neumann from Los Alamos on 15 December 1945.[62] The war was over, and Klári visited the secret lab for the first time over Christmas. The curtains of secrecy that had hung over von Neumann's life were lifted at last. Klári was welcomed into the tight-knit community of scientists, technicians and their families. Von Neumann had not lied: there were excellent prospects for skating on Ashley Pond, named after the founder of the Ranch school, which froze solid during the winter months. The parties, the late-night poker sessions and the many Hungarians there must also have reminded her of the charmed life she had led in Budapest. There in the desert, she found again many of the things she had missed during her isolated existence in Princeton – including her husband. She also began to learn something about Teller's Super, and the epic scale of the problem facing the mathematicians and physicists charged with finding out if the bomb was workable. Teller had concluded the task was beyond mechanical calculators or IBM's punched-card machines. Under von Neumann's direction, they had begun modelling the ignition process on the computer built in secret at the University of Pennsylvania's Moore School. Von Neumann's 'obscene interest' in computing machines had by now turned into an obsession. Klári caught the bug. With her husband's encouragement she began familiarizing herself with the ENIAC. In the summer of 1947, Los Alamos hired her as a consultant. 'I learned how to translate algebraic equations into numerical forms,' she reflected many years later, 'which in turn then have to be put into machine language in the order in which the machine was to calculate it.' In other words, she 'became one of the first "coders"'. She described her new occupation in terms that would be familiar to many programmers today. It was, she said, a 'very amusing and rather intricate jig-saw puzzle' that was 'lots and lots of fun'.[63]

Von Neumann and Ulam had in the meantime developed a new technique that harnessed the laws of chance to provide approximate answers to equations that could not be solved exactly by traditional means. Klári's task would be to bring the ENIAC's number-crunching power to bear on the mathematics of neutrons diffusing inside a nuclear weapon, which was precisely a problem of this sort. She must have been tickled to hear that the new technique was coincidentally named the 'Monte Carlo' method, after the town where she and Johnny had first met.

The idea of exploiting the power of randomness to solve problems had come to Ulam while convalescing from viral encephalitis in a hospital bed. Told by his doctors to rest his inflamed brain, Ulam began to play solitaire to relieve the boredom.[64] Unable to keep his mental activity in check, Ulam began to calculate the probability of playing out a hand of solitaire successfully to completion.[65] The numbers quickly got out of hand; there were too many possible card combinations. 'After spending a lot of time trying to estimate them by pure combinatorial calculations, I wondered whether a more practical method than "abstract thinking" might not be to lay it out say one hundred times and simply observe and count the number of successful plays.'

Ulam realized that many real-world problems are surprisingly similar in nature to working out the chances of winning solitaire. A complex situation can be made tractable by setting up a model that is then run repeatedly to reveal the most likely outcomes. 'It's infinitely cheaper to imitate a physical process in a computer and make experiments on paper, as it were, rather than reality,' Ulam explained later.[66]

Monte Carlo made simulating a chain reaction possible for the first time. There are a vast number of different ways in which an assembly of neutrons might behave – too many to calculate. But by plotting the random path of lone neutrons hundreds of times over, one can build up an accurate picture of the reaction – exactly the sort of analysis that Los Alamos urgently needed to improve the efficiency of bombs.

Ulam presented his idea to von Neumann during his next visit to Los Alamos. Von Neumann was preparing to leave, so Ulam hopped into the government car taking him to the train station, and they fleshed out the details together during the long journey. In March 1947, von Neumann sent an eleven-page plan for running Monte Carlo bomb simulations on an electronic computer to Robert Richtmyer, head of the Los Alamos theoretical division. Computers now run Monte Carlo simulations thousands of times a day, and applications range from optimizing stock portfolios to testing the properties of new materials.

Von Neumann described what was essentially a simplified atom bomb comprising a series of concentric shells. The composition of each shell (chosen to reflect the properties of actual metals and alloys) would determine the chances that neutrons would slow down, trigger fission, be absorbed or be reflected back towards the core. Von Neumann proposed plotting the path of a hundred neutrons through the bomb by using random numbers to select the outcome of every interaction. He also included a list of eighty-one steps in a computation that would take a single neutron through one cycle of the Monte Carlo process.

Over the next few months, von Neumann and Goldstine wrote reports setting out their approach to programming. They introduced flowchart diagrams – still used to represent computer algorithms – to plan the Monte Carlo program. An office at the IAS became a hub for planning the computer simulation and became known as the 'Princeton Annex' of Los Alamos. Adele Goldstine and Richtmyer were the first occupants but they soon left to prepare other fission calculations, code-named 'Project Hippo'. From then on, Klári was in charge. She set to work translating von Neumann's flowcharts into workable computer code.[67]

Klári and Johnny arrived in Aberdeen on 8 April 1948, but he left soon afterwards. Metropolis had arrived a couple of weeks earlier, ready to run the Monte Carlo program that Klári had developed. But the ENIAC was not ready. The BRL staff had been hard at work for many months planning the conversion to the EDVAC-style 'stored-program' mode, but though the plans had been reported in the *New York Times* in December 1947,[68] the reconfiguration had not begun.

The planned fifty-one 'hard-wired' instructions had been expanded to sixty. With Adele now at work on the Hippo calculations, Bartik and her team had taken on the task with Richard Clippinger, a mathematician who hoped to use the ENIAC to simulate supersonic airflows. Bartik's team had worked out how to rewire the ENIAC for the expanded instruction set and wrote a number of programs to run once the conversion was complete.

In the end, Metropolis and Klári gave the ENIAC an even more extensive vocabulary, with seventy-nine possible commands, using the older plans of Adele and the BRL team as templates. Three weeks later, the ENIAC was ready. The machine was now, in effect, a stored-program computer – the first of the millions more to come. 'With the help of Klári von Neumann, plans were revised and completed and we undertook to implement them on the ENIAC,' says Metropolis.

'Our set of problems – the first Monte Carlos – were run in the new mode.'[69]

In his letter to Richtmyer, von Neumann had broached the idea of using a punched card to represent a single neutron at one moment in time. The card would also carry a pre-generated random number that would determine the neutron's fate. After the neutron was absorbed, scattered or fissioned, a new card would be produced and manually fed back into the machine. By December 1947, that approach had been abandoned. Klári's program used the ENIAC's new computing capacity to accelerate the simulation.

A neutron's journey inside the bomb was followed for a full ten nanoseconds (an interval known as a 'shake' in nuclear physics), which often meant that it was scattered several times before a new card had to be punched. Von Neumann had also come up with a way of producing pseudo-random numbers internally by squaring an eight- or ten-digit binary number and using the middle digits (which could also be squared to produce the next 'random' number and so on). The numbers produced with von Neumann's 'method of middle-squares' were not truly random, as he knew. 'Any one who considers arithmetical methods of producing random digits is, of course, in a state of sin,' he noted later.[70] But they were good enough for his purposes and, it turned out, for many others too.

The calculations began on 28 April, and by 10 May the first Monte Carlo run was over. 'I heard from Nick on the telephone that the miracle of the ENIAC really took place,' Ulam wrote to von Neumann shortly afterwards, 'and that 25,000 cards were produced!!'[71]

Time on the ENIAC was precious, and the team laboured all hours of the day and night to get the job done. 'Klári is very run-down after the siege in Aberdeen, lost 15 lbs.,' von Neumann told Ulam.[72] Plans for a joint holiday with the Ulams were dropped, and Klári had a check-up at Princeton Hospital, complaining a month later that she 'was still being annoyed by various tests and treatments'.[73] Still, she was able to write up a report on the conversion and use of the ENIAC. Expanded and edited, the document was to become the definitive record of the Monte Carlo runs she and Metropolis had overseen.

Klári returned to the BRL in October. Historians have recently recovered the complete program for this second Monte Carlo run: twenty-eight pages of code written in Klári's hand.[74] The calculations were completed on 7 November. 'Klari survived the Aberdeen expedition this time better than the last one,' von Neumann wrote to Ulam.[75] A solo trip to Los Alamos followed in December, but the stress of defending her work before the likes of Teller and Fermi nearly tipped Klári over the edge. Finding her 'catastrophically depressed' during a telephone call, von Neumann wrote to her, professing himself 'scared out of my wits' with worry. Nonetheless, less than six months later, she travelled to Chicago to work with theoretical physicist Maria Goeppert Mayer, a future Nobel laureate, on possible refinements to the Monte Carlo algorithm. Mayer's ideas for improving the calculations were ultimately rejected. With little time left on the ENIAC to run the Monte Carlo simulations that Los Alamos wanted, von Neumann opted to play it safe and avoid new, and potentially troublesome, refinements.

'Things are kind of upside down,' Klári wrote to the Ulams shortly before the third Monte Carlo run began. 'Please pray for me and hope for the best.'[76] The last of the Monte Carlo problems were finished, successfully, on 24 June 1949. Klári returned home exhausted on 28 June, reporting she had 'brought with me to Princeton all secret documents' – probably nuclear cross-section data showing the chances of fission occurring in different materials. Ten large boxes of punched cards were despatched to Los Alamos. On 7 July, Klári jumped on a flight to follow them, ready once more to review the calculations before the cream of the world's physicists and mathematicians. In 1950, she returned to Aberdeen for the last time. On this occasion, the job was to put Teller's Super design to the test, the question being whether the fission bomb would be able to trigger the more powerful fusion device. The simulations confirmed that Teller's assembly would not generate enough heat to do so, and the design was finally abandoned. The 'staged implosion' hydrogen bombs that Teller and Ulam later devised would have a very different method of ignition.

After the 1950 Super calculations, Klári retired from the forefront of computing. Los Alamos soon had its own computer, whimsically named MANIAC I and built by a team under the direction of Metropolis.[77] Her expertise was still sought out: 'Would you look carefully to see that we haven't done anything completely outrageous?' asks one letter from Los Alamos, as the team there prepared to run more bomb simulations. But plagued by insecurity and worsening bouts of depression, she penned no further code herself, even when her husband's machine at the IAS started running reliably in 1952.

The full scope of Klári's contributions to the early days of computing have only recently come to light. Run on an ENIAC emulator today, her Monte Carlo code reliably spits out the expected numbers, virtually plotting the fates of neutrons inside the implosion bomb that von Neumann had helped build. The world's first electronic stored-program computer is usually acknowledged to be the Small-Scale Experimental Machine (SSEM) built at the University of Manchester in England. Known as the 'Manchester Baby', the computer ran its first program on 21 June 1948, two months after Klári's code ran on the reconfigured ENIAC in Aberdeen.[78] The Manchester Baby cycled through seventeen instructions over fifty-two minutes to determine that the highest factor of 262,144 is 131,072, a calculation that a bright primary-school child could do in considerably less time. Klári's 800-command program that ran in April in Aberdeen was used to adjust the composition of atom bombs. Within that program is a 'closed subroutine' – a type of loop that is executed whenever it is referenced from the main body of the program. The invention of the closed subroutine is generally credited to computer scientist David Wheeler, but Klári's code made use of one at least a year earlier, to generate random numbers by von Neumann's 'method of middle-squares'.

Debate still rages in some quarters over whether the ENIAC in its new guise really constituted a true 'stored-program' computer. There can be little doubt, however, that Klári's Monte Carlo code is the first truly useful, complex modern program ever to have been executed.

The much-delayed IAS machine finally roared into life in 1951. Chief engineer Bigelow was something of a perfectionist. He and Goldstine disagreed on almost every important decision. Only von Neumann could keep things ticking along. 'He kept Herman and I from fighting by some marvellous technique,' Bigelow remembered. 'We got along like oil and water, or cat and dog; and von Neumann would keep this here, and this there, and smooth things over.'[79]

The IAS computer in 1952. *Alan Richards photographer. From the Shelby White and Leon Levy Archives Center, Institute for Advanced Study, Princeton.*

When Bigelow was awarded a year-long Guggenheim fellowship, Goldstine quickly appointed fellow electrical engineer James Pomerene to temporarily replace him. The computer, which had been close to completion for some time, was finished in Bigelow's absence, though built according to his design. The writer George Dyson memorably compares the machine's appearance to 'a turbocharged V-40 engine, about 6 feet high, 2 feet wide, and 8 feet long'. 'The computer itself,' he continues, 'framed in aluminium, weighed only 1,000 pounds, a microprocessor for its time. The crankcase had 20 cylinders on each side, each containing, in place of a piston, a 1,024-bit memory tube.'[80]

By this time, the IAS machine's numerous offspring, built with the aid of Goldstine and von Neumann's numerous progress reports, were snapping at its heels. Metropolis's MANIAC I went into operation at Los Alamos in 1952. The national laboratories at Oak Ridge and Argonne unveiled, respectively, the ORACLE and the AVIDAC the following year. Most important of all, however, was the IBM 701, unveiled to the public in 1953 and the company's first commercial computer.

A machine made for scientific research, the IBM 701 marked a turning point for a firm that was still making much of its money by selling machines descended from punched-card tabulators. Bigelow had worked for IBM for several years before joining von Neumann's project at the IAS. In 1938, he says, 'IBM was a very mechanically-orientated company and the notion of electronic computing was almost repugnant.'[81] Spurred by von Neumann and the numerous computers springing up in the wake of his project, the company rapidly changed course, producing digital stored-program machines in the EDVAC mould. The IBM 701 was, says Bigelow, 'a carbon copy of our machine'.[82] By the 1960s, IBM manufactured about 70 per cent of the world's electronic computers. 'Probably', Teller told his biographers, 'the IBM company owes half its money to Johnny von Neumann.'[83]

Did von Neumann understand the potential of the machines he helped to invent? Yes, he did. In reflective mood in 1955, he noted that the 'over-all capacity' of computers had 'nearly doubled every year' since 1945[84] and often implied in conversation that he expected that trend to continue. His observations prefigure 'Moore's law', named after Intel's cofounder Gordon Moore, who predicted in 1965 that the number of components on an integrated circuit would double every year.

By a quirk of history, the individual who had perhaps the deepest understanding of the logical and mathematical underpinnings of the modern computer also had the power, influence and managerial skills to build one, as well as the good sense to ensure that the race to faster, more powerful machines he initiated was run (at least to begin with) in public. 'Von Neumann cleared the cobwebs from our minds as nobody else could have done,' wrote Bigelow long afterwards. 'A tidal wave of computational power was about to break and inundate everything in science and much elsewhere, and things would never be the same.'[85]

6

A Theory of Games

Perking up the dismal science

> *'Omar Little: I got the shotgun. You got the briefcase. It's all in the game, though, right?'*
>
> *From* The Wire, *2003*

Von Neumann was rational. At times some may have thought him *too* rational. Perhaps this is best illustrated by the child custody arrangements that he and his first wife Mariette came up with for their daughter, Marina, when she was just two years old. The two agreed that until the age of twelve, Marina would live with her mother and spend holidays with her father. After that, when she was 'approaching the age of reason', Marina would live with her father to receive the benefit of his genius.[1]

'It was a thoughtful and well-intentioned agreement,' Marina says in her memoirs, 'but they were too inexperienced to realize that adolescence is often the stage in life farthest removed from the age of reason.'[2]

Her father's letters, Marina notes, often reflected his 'lifelong desire to impose order and rationality on an inherently disorderly and irrational world'.[3] Game theory sprang from von Neumann's urge to find neat mathematical solutions to knotty real-world problems during one of the most 'disorderly and irrational' periods in human history. The answers of game theory sometimes seem cold, unconventional, shorn of the complexities of human emotion – but effective, nonetheless. Marina, incidentally, grew up to become a noted economist, and the first woman to serve on the President's Council of

Economic Advisers, so perhaps the same could be said of the von Neumanns' custody arrangements.

What is game theory about? Not what you might expect. Von Neumann once mentioned the term to the mathematician Jacob Bronowski in a London cab during the war. Bronowski wrote about the conversation in *The Ascent of Man*:

> And I naturally said to him, since I am an enthusiastic chess player, 'You mean, the theory of games like chess?' 'No, no,' he said. 'Chess is not a game. Chess is a well-defined form of computation. You may not be able to work out the answers, but in theory there must be a solution, a right procedure in any position. Now real games,' he said, 'are not like that at all. Real life is not like that. Real life consists of bluffing, of little tactics of deception, of asking yourself what is the other man going to think I mean to do. And that is what games are about in my theory.[4]

Von Neumann was not the first to analyse conflicts in this way. Though chess does not feature much in modern game theory textbooks for the reasons he described, it was the game of kings that first seems to have inspired mathematicians in Germany and Austria-Hungary to theorize about the psychology of conflict. Von Neumann was among them. He may not have been an exceptional player but when he moved to Zurich in 1925, he immediately joined the city's famous chess club, one of the oldest in the world.[5]

Foremost among the great chess strategists was the legendary Prussian player Emanuel Lasker, who was the world champion for twenty-seven years from 1894. Lasker's first love was mathematics, which he studied at Berlin, Heidelberg and Göttingen, where David Hilbert took him under his wing. Despite Hilbert's patronage and a string of impressive papers, Lasker struggled to get a permanent position in Germany because he was Jewish. After temporary lectureships in Manchester and New Orleans, he began to earn a living by playing chess and pursued mathematics in his spare time.

Lasker became famous for shunning textbook moves, making risky plays and throwing his opponents into confusion. Psychology was central to his game: Lasker did not play chess, it was said, so much as the man in front of him.[6] 'He who relies solely upon tactics that he can wholly comprehend', he warned, 'is liable, in the course of time, to weaken his imagination.'[7]

Lasker drew parallels between chess, war and the struggles of economic and social life. The task of the perfect strategist, he argued, is to gain the upper hand with as little effort as possible.[8] In the closing chapter of his *Manual of Chess*, published in 1925, Lasker expressed his desire for a new mathematics for resolving conflicts. His mentor, Hilbert, was at the height of his powers, and the dream of axiomatizing mathematics and the sciences was still alive in the hearts of his followers. Why not, Lasker reasoned, also a rigorous theory of human cooperation and dissent?

'The science of contest', Lasker predicted, 'will progress irresistibly, as soon as its first modest success has been scored.' Institutes dedicated to the new discipline would 'breed teachers capable of elevating the multitude from its terrible dilettantism' in matters of negotiation, transforming politics completely, and 'aid the progress and the happiness of all humankind'. Ultimately, he writes, their aim would be to render war obsolete by providing rational methods for reaching agreements.[9] Some have suggested game theory was the product of cynical minds. One could argue that, on the contrary, at the root of the discipline was the naive hope that mathematics could help forge lasting peace.

In 1925, the tools to realize Lasker's grandiose ambitions were not available. The first decisive step towards a 'science of contest' would be taken towards the end of the following year. On 7 December 1926, von Neumann unveiled his proof of the minimax theorem to mathematicians at Göttingen. Published in 1928, the paper expounding his proof, *On the Theory of Parlour Games*,[10] would firmly establish game theory as a discipline, framing human cooperation and conflict in truly mathematical terms.

Von Neumann had been interested in the scientific principles underlying games and toys since childhood. Among the knickknacks on his desk were a glass tube filled with soap bubbles and a spinner. 'Watching the changing pattern of the soap bubbles after he shook the glass tube, he contemplated the effect of surface tension in making them obey the law of entropy; noting where the pointer on the wooden disk landed on spin after spin stimulated his ideas about the laws of probability,' says Marina. 'Had LEGOs been available at the time, he might have built a model of his computer from them.'[11]

Von Neumann had already helped some of his Hungarian colleagues with their papers on the mathematics of chess strategy. As one of Hilbert's protégés he was still under the great mathematician's spell and harboured the dream of extending the axiomatic method to fuzzy sciences like psychology. He was naturally drawn to the mathematical study of games.

Von Neumann begins his minimax paper by stripping a game of strategy to its bare essentials. Consider a card game such as rummy. Each player is dealt a hand and on each round must decide what cards to play. That decision is based on the cards they hold and whatever cards have been laid down by any preceding players. If they can't play, they draw from the deck, and the turn passes to the next player. The game ends when a player has managed to discard all their hand as melds, at which point the losing players score the cards remaining in their hands.

Losing none of the vital aspects of the game (except the fun), this can be boiled down to a series of events, some of which depend entirely on chance (von Neumann calls these 'draws') and others which 'depend on the free decision' of the players (he calls these events 'steps'). At the end of the game, each of the players receive a payout (points or money), which is dependent on the outcome of all draws and steps. 'How must one of the participants . . . play in order to achieve a most advantageous result?' von Neumann asks. 'We shall try to investigate the effects which the players have on each other, the consequences of the fact (so typical of all social happenings!) that each player influences the results of all other players, even though he is only interested in his own.'

No player has complete control over their payout because they can neither determine the results of the 'draws' nor the 'steps' taken by other participants. Von Neumann demonstrates that in mathematical terms the 'draws' do not matter: the effects of chance can be accounted for by modifying the strategies available and the size of the players' payouts. Ultimately, a game can be represented simply as the choice by each player of a single strategy (effectively an amalgam of all the strategies they play in the game), followed by a calculation of their respective payouts that accounts for everyone's choices (and which factors in their luck).

Von Neumann could not get any further with multi-player games, so he switched to thinking about a situation with just two opponents whose individual payouts sum to zero. 'It is not enough to succeed. Others must fail,' Iris Murdoch once wrote. Von Neumann coined the term 'zero-sum' to describe such games of total conflict, in which one person's loss is the other's gain. One indication of the influence of game theory is that 'zero-sum' has now passed into the vernacular.

In von Neumann's paper, the players, S_1 and S_2, each choose a strategy, x and y, without knowing what the other has chosen. At the end of the game, player S_1 gains an amount g, which depends on x and y. Conversely, player S_2 loses the same amount, g, because the game is zero-sum. Von Neumann imagined a tug-of-war between the two sides: 'It is easy to picture the forces struggling with each other in such a two-person game,' he says. 'The value of $g(x,y)$ is being tugged at from two sides, by S_1 who wants to maximize it, and by S_2 who wants to minimize it. S_1 controls the variable x, S_2 the variable y. What will happen?'[12]

Imagine that we bend the rules of the game and allow the two players to know in advance the strategy their opponent will play. Von Neumann's game theory assumes that all players are perfectly rational.[13] The most rational strategy for S_2 to play is the one which ensures player S_1's payoff is as small as possible. But player S_1 is brutally rational too and, knowing that S_2 is a grasping individual, S_1 will play the strategy guaranteed to secure her the maximum possible payoff if, as she expects, S_2 will play to cut any gains from her strategy to the bone. Player S_2's effort to minimize his maximum loss is the 'minimax' strategy. It is the best player S_2 can do in a somewhat paranoiac world where it is certain that the other player will assume the worst about you. The best player S_1 can do is known as the 'maximin' strategy, which will secure her the maximum possible payoff, assuming her opponent will not willingly lose any more than he absolutely must. Von Neumann set out to prove that every two-player zero-sum game similarly has a 'solution'. That is, a strategy for each player that guarantees the best outcome *given* they are up against a rational player who is exclusively out for themselves too.

A very simple illustration of a game of this sort is the classic cake-cutting conundrum: a hapless parent facing the prospect of dividing a cake between two rival siblings decides to let one sibling cut the cake and the other choose the piece they want. The first sibling, if she is behaving as rationally as game theory requires, will pursue the maximin strategy of cutting the cake as evenly as possible – she predicts her sibling will choose the bigger piece, so she wants to maximize the minimum amount of cake she'll be left with. The second sibling's minimax strategy is (in this elementary example) obvious: choose the bigger piece.

In this game, the minimax and maximin coincide: the best possible outcomes for both the cutter and chooser in the face of a smart, selfish sibling are the same half-cake, give or take a few crumbs. This is called a 'saddle point', akin to a mountain pass that is both the highest point reached by the path cutting through the range and the lowest point of the valley rising to either side.

Most two-person, zero-sum games are far more complex than cake-cutting, and many do not have saddle points. What von Neumann was proposing to prove was far from self-evident. In fact, another early pioneer of game theory, the French mathematician Émile Borel, had concluded that there was no general solution to the problem.[14] Borel wrote several papers on game theory in the early 1920s, and his definition of the 'best strategy' was the same as von Neumann's – the choice made by a perfectly rational player who wants to win or lose as much or as little as possible. Despite demonstrating that solutions existed in two-person games where there were only three or five strategies to choose from, Borel was sceptical that there could be optimal solutions in all cases.

Borel's papers on the subject were not widely known until 1953, when his compatriot Maurice Fréchet had them translated into English by mathematician Leonard Savage.[15] In an accompanying commentary, Fréchet argued Borel was the 'initiator' of the field.[16]

When von Neumann learned of the claim, he was furious. 'He phoned me from someplace like Los Alamos, very angry,' Savage recalled. 'He wrote a criticism of these papers in English. The criticism was not angry. It was characteristic of him that the criticism was written with good manners.' Von Neumann, who would never forget being beaten to the punch by Birkhoff and Gödel, was in no mood to give up his claim to being game theory's inventor. He had not seen Borel's papers when he was formulating his proof, von Neumann said in his response. If he had, the effect would have been 'primarily discouraging'.[17] 'Throughout the period in question I thought there was nothing worth publishing until the "minimax theorem" was proved,' he added. 'As far as I can see, there could be no theory of games on these bases without that theorem. By surmising, as he did, the incorrectness of that theorem, Borel actually surmised the impossibility of the theory as we now know it.'

Outside France, von Neumann's 1928 minimax paper is today recognized as the founding work of game theory. The key to that proof are 'mixed strategies', a concept he later illustrated with another simple two-person zero-sum game called 'Matching Pennies'. Two players each with a penny secretly turn the coin to show either heads or tails. One player wins both pennies if they match; the other if they do not. There is no saddle point: you either win or lose a penny and there is no 'best strategy' anyone can adopt before knowing the other person's choice. Putting aside the objection that anyone actually forced to play the game would be quickly overwhelmed by boredom, there is an obvious 'winning' strategy for Matching Pennies: play heads or tails at random. This 'mixed strategy' allows players to break even in the long run. A 'pure strategy' of picking either heads or tails consistently invites a canny opponent to take advantage of your naivety.

Israel Halperin, von Neumann's only doctoral student, called him a 'magician'. 'He took what was given and simply forced the conclusions logically out of it, whether it was algebra, geometry, or whatever,' Halperin said. 'He had some way of forcing out the results that made him different from the rest of the people.'[18] Hungarian mathematician Rózsa Péter's assessment of his powers is more unsettling. 'Other mathematicians prove what they can,' she declared, 'von Neumann proves what he wants.'[19]

Von Neumann's minimax proof is in this vein; he bulldozes his way through six pages of dense algebra to reach his conclusion that every two-person zero-sum game has a solution that is either a pure or mixed strategy.

Many familiar games require a player to use a mixed strategy to win. As every cardsharp knows, an element of unpredictability keeps opponents off-balance. Von Neumann, recognizing the implications of his work, finishes his paper with a flourish: 'The agreement of the mathematical results with the empirically known rules of successful gambling, for example, the necessity of bluffing in poker, can be considered as experimental confirmation of our theory.'

Game theory is now best known as a branch of economics but von Neumann only considers the connection briefly in his 1928 paper. 'Any event,' von Neumann says, 'may be regarded as a game of strategy if one looks at the effect it has on the participants.' He describes this interplay between players as the 'principal problem of classical economics'. 'How,' he muses, 'is the absolutely selfish "homo economicus" going to act under given external circumstances?'

Von Neumann would not return to this question for more than a decade. There would be no major developments in game theory until he did. In the meantime, he found other economic avenues to explore. In 1932, von Neumann gave a half-hour seminar at Princeton entitled 'On Certain Equations of Economics and a Generalization of Brouwer's Fixed-Point Theorem'. He delivered the unscripted talk in German, and though most present could be expected to have understood him, von Neumann's machine-gun delivery accompanied by his usual habit of wiping the blackboard clean before anyone could catch up with him probably explained why his thoughts did not immediately reach a wider audience. He was asked to deliver the seminar again in Vienna four years later, and this time he wrote it up. The dense, nine-page paper was published in 1937 as part of the proceedings of the colloquium.[20] In 1945, his friend the left-wing Hungarian economist Nicholas Kaldor (born Miklós Káldor and later Baron Kaldor) arranged for the paper to be translated into English.[21] 'Johnny agreed to this,' said Kaldor. 'Indeed he was grateful for any effort that would enable his model to reach a wider audience as he was much too preoccupied with whatever he was working on at the moment (I think it may have been the computer).'[22] Von Neumann returned his corrections from Los Alamos. He was, as Kaldor had guessed, working on 'something else'.

Von Neumann's 'Expanding Economy Model', as it is now known, shows that an economy will 'naturally' reach a maximum growth rate – a 'dynamic equilibrium' driven by cycles of production, consumption and decay. In von Neumann's model, when the equilibrium is reached, all goods are produced as cheaply and as abundantly as possible. Older models had tended to assume that an equilibrium existed; von Neumann proved that one arose as a direct consequence of his axioms, including the assumption that an infinite supply of labour was available, for instance, and 'all income in excess of necessities of life will be reinvested'.

Von Neumann's proof exploited a powerful result from the field of topology, discovered by Bertus Brouwer, the same mathematician who had irked Hilbert so with 'intuitionism'. Brouwer's 'fixed-point theorem' states that, for certain mathematical functions, at least one result output by the function is the same as the number that is put in at the start. Plot the function on a graph, and at these special 'fixed points' the x and y coordinates will be the same.[23]

One way to visualize Brouwer's topological proof of this is to imagine superposing two maps with different scales that both depict the same place. As long as the smaller map is placed inside the larger one, there will be at least one point where a pin pushed through both would pierce the same geographical point – no matter what the relative orientations of the two maps.

Von Neumann stressed that his model was not a detailed simulation of a real economy but a crude metaphor. He felt the discipline of economics was not ready for a more concrete treatment. 'Economics, as a science, is only a few hundred years old,' he explained later.

> The natural sciences were more than a millennium old when the first really important progress was made . . . methods in economic science are not worse than they were in other fields. But we will still require a great deal of research to develop the essential concepts – the really usable ideas.[24]

Privately, he was less diplomatic. 'If these books are unearthed sometime a few hundred years hence, people will not believe that they were written in our time,' von Neumann confided to a friend in 1947, referring to some of the discipline's most lauded contemporary works.

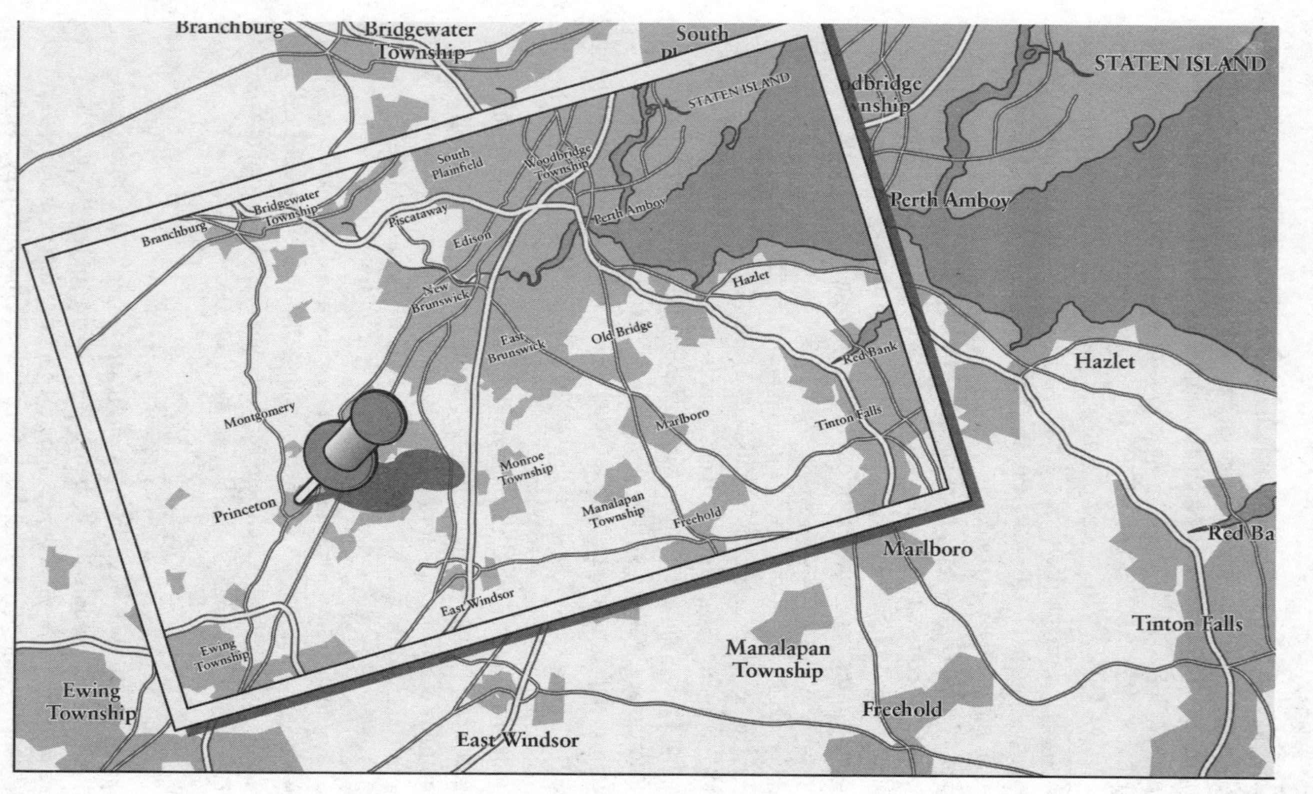

Brouwer's fixed-point theorem. Imagine laying one map on top of another larger map of the same place. One point will always overlap.

'Rather they will think that they are about contemporary with Newton, so primitive is their mathematics. Economics is simply still a million miles away from the state in which an advanced science is, such as physics.'[25]

Predictably, most economists found von Neumann's translated paper impenetrable. The old guard regarded his use of a topological theorem to prove the central result a bizarre mathematical affectation. Some thought von Neumann was advocating a slave economy, with endless labour available for subsistence-level pay. He was not – though one implication of his model is that precipitous wage rises will retard economic growth and depress wages in the longer term. Others criticized some of the model's unrealistic premises. The English mathematical economist David Champernowne, who *did* understand the paper, praised 'the elegance of the mathematical solution of a highly generalised problem in theoretical economics' but noted the 'extremely artificial assumptions' that were necessary to derive the result.

Despite the generally frosty reception, 'A Model of General Economic Equilibrium' sparked a revolution. Mathematicians, inspired by von Neumann's achievement, poured into economics and began applying fresh methods to the dismal science. By the 1950s, the subject was transformed. Fixed-point theorems were used to prove key results in economics – including in von Neumann's own game theory by a young upstart called John Nash. A half-dozen Nobel laureates are reckoned to have been influenced by the work.[26] Among them were Kenneth Arrow and Gérard Debreu, who were awarded the prize (in 1972 and 1983 respectively) for their work on the theory of general equilibrium, which models the workings of a free-market economy. A half-century after von Neumann's Princeton seminar, the historian Roy Weintraub described his paper as 'the single most important article in mathematical economics'.[27]

But as was his way, von Neumann had moved on long before anyone really recognized its significance. A year before the translated paper appeared, he produced *Theory of Games and Economic Behavior*, the book that would forever change the social sciences and profoundly influence economic and political decision-making from the 1950s to the present day. What made him return to game theory after his minimax proof of 1928 was the encouragement of a new friend – the German economist Oskar Morgenstern.

Morgenstern was an oddball. Tall and imperious, he rode through Princeton on horseback, immaculately attired in a business suit. He was born on 24 January 1902 in Görlitz, then in the Prussian province of Silesia, but grew up in Austria. His mother was the illegitimate daughter of German Emperor Frederick III and after he moved to the United States, Morgenstern proudly kept a portrait of his grandfather, the Kaiser, hanging in his home.[28]

At school, he shared the nationalist passions of his peers. 'What would the world be without the German people and its achievements!' he wrote to a friend, adding more sinisterly, 'All the foreign dirt which will wipe out healthy ideas, has to be removed with violence.'[29]

Morgenstern was not a very able mathematician and he learned no more of the subject during his degree in political science at the University of Vienna. He was initially drawn into the orbit of conservative thinker Othmar Spann, whose lectures were wildly popular with students. Spann argued that in a capitalist democracy the downtrodden masses would inevitably vote in socialist regimes. His solution was a strong, authoritarian state, a conclusion eagerly embraced by the nascent fascist parties of Europe. Spann sang the praises of Hitler and eventually joined the Nazi Party.

Eventually, Spann's rhetoric lost its hold on Morgenstern, who discovered principally through the Jewish economist Ludwig von Mises the classical liberalism of the Austrian School. Mises's belief in the power of free markets put him at odds with other Jewish professors, who were mostly left-leaning progressives. Without their support, Mises was unable to overcome the general antipathy of the gentile faculty and never secured a tenured position. He was still extraordinarily influential.

Among those who flocked to the private seminars Mises held in his office was future Nobel laureate Friedrich Hayek, whose criticisms of central planning and socialism would inspire economic liberalizers like Margret Thatcher, Ronald Reagan – and Chilean dictator Augusto Pinochet. Morgenstern attended the 'Mises Seminars' for many years, though he complained in his diary that he was 'uncomfortable' as 'the only pure Aryan (out of 8!)' and deplored the 'unpleasant discussion in this arrogant circle of Jews'.[30]

In 1925, the twenty-three-year-old Morgenstern submitted his doctoral thesis. He had impressed the professors at the university sufficiently that they helped him secure a Rockefeller-funded fellowship, and over the next three years Morgenstern would travel to Britain, the United States, France and Italy. In England, he was most impressed by the statistician Francis Edgeworth, who had recently retired as the Drummond Professor of Political Economy at Oxford University. As a result, after many years of indifference, Morgenstern developed a newfound respect for mathematics. 'In Germany,' Morgenstern wrote after Edgeworth's death the following year, 'every beginner believes . . . that he should create entirely "new" foundations and a completely new methodology. In England, however, the use of mathematics is more common than elsewhere, a further circumstance to frighten away dilettantes who are only half interested in economics.'

During his fellowship, Morgenstern developed an interest in the ebb and flow of business cycles. After stints with experts at Harvard and Columbia, Morgenstern made the subject the focus of his *habilitation* thesis, which would earn him the right to teach and become a tenured professor. The professors in America had sought to explain booms and busts by careful analysis of economic statistics with the eventual goal of anticipating downturns. But what Morgenstern had learned about mathematics made him pessimistic. His dissertation was a sustained attack on economic forecasting, which he argued was impossible. When his screed was published in 1928, contemporary reviews were hostile. 'One feels', said one, 'that Dr. Morgenstern has here given us what, despite his declared intention, can be described only as a satire.' After the Wall Street crash the following year, Morgenstern's stock rose as share prices tumbled.

The crux of Morgenstern's argument was that any prediction would be acted on by businesses and by the general public, and their collective responses would invalidate it. Any updated forecast would be dogged by the same problem. Morgenstern recognized this cycle of guess and counter-guess at work in 'The Final Problem', a Sherlock Holmes story, in which Arthur Conan Doyle's shrewd detective is being pursued by his mortal enemy, Professor James Moriarty, who will certainly kill Holmes if he can catch him:

> An analogy would be appropriate here, which is at the same time quite amusing: as Sherlock Holmes, chased by his enemy Moriarty, leaves from London to Dover with a train which stops at an intermediate station, he gets off the train instead of going on to Dover. He saw Moriarty at the train station and, considering him very intelligent, expected that Moriarty would take the faster train, to await him in Dover. This anticipation of Holmes turned out to be right. But what if Moriarty had been even more intelligent and had considered Holmes' capacities even greater, and therefore had predicted Holmes' action? Then he [Moriarty] would obviously only have gone to the intermediate station. Again Holmes would have calculated that, and therefore would have chosen Dover. Thus, Moriarty would have acted differently. And from so much thinking, there would have come no action, or the less intelligent one would have handed himself to the other at Victoria Station because all the fleeing would have been unnecessary. Examples of that kind could be taken from everywhere. Chess, strategy, etc., but there one needs to have special knowledge, which simply makes the examples more difficult.[31]

Morgenstern continued to write about this conundrum, criticizing other economists for failing to account for this infinite regress in their theories. But he got no further with tackling the problem himself. In the mid-1930s, when he was presenting related work, a mathematician told him that a certain John von Neumann had dealt with exactly this issue in a 1928 paper on the theory of games.[32] Morgenstern's interest was piqued but not enough to wade through von Neumann's minimax proof. He was by this time director of the Austrian Trade Cycle Institute in Vienna, a post he had shared with Hayek from 1928 until 1931, when Hayek moved to the London School of Economics.

In January 1938, Morgenstern took a leave of absence to lecture in the United States at the invitation of the Carnegie Endowment for International Peace. He left his deputy, Reinhard Kamitz, in charge. On 12 March, German troops marched into Austria, and Kamitz (who would serve as an Austrian finance minister after the war) cheerfully appeared at the institute in full Nazi regalia. Morgenstern was deemed 'politically unbearable' and summarily dismissed from his job. The *Anschluss* appears to have cured Morgenstern of his anti-Semitism. The snide jibes aimed at Jews largely disappear from his diaries after 1938. His own family in Austria were forced to prove their Aryan ancestry (which they did, back to the 1500s).

Informed that, probably thanks to Kamitz, he was on the Gestapo's blacklist, Morgenstern decided not to return to Austria. He received job offers from a number of universities but accepted a lectureship at Princeton. This was partly in the hope of meeting von Neumann. Mostly, he wanted to use the post as a beachhead to schmooze his way to 'the double income etc. and no worries'[33] of a professorship at the nearby IAS. 'If I only had a position at the Institute,' he grumbled.[34]

Morgenstern found Princeton to be much as Einstein had described it five years earlier; 'a quaint and ceremonious village of puny demigods on stilts'.[35] He soon scandalized polite society by marrying a

Oskar Morgenstern. *Courtesy of the University of Vienna archives.*

beautiful redheaded bank teller named Dorothy Young, some fifteen years younger than he was. Soon after the university term began that year, he met von Neumann.

'It is curious,' Morgenstern noted,

> that years later neither of us could ever remember where we met the first time, but we did remember where we met the second time: I gave an after-luncheon talk on the 1st of February 1939 on business cycles at the Nassau Club, and he was there with Niels Bohr, Oswald Veblen, and others. Both he and Bohr invited me that afternoon for tea at Fine Hall, and we sat several hours talking about games and experiments. This was the first time that we had a talk on games, and the occasion was heightened by Bohr's presence. The disturbance of experiments by the observer was, of course, one of the famous problems raised by Niels Bohr for quantum mechanics.[36]

Bohr was drawing a parallel between the perturbations caused by the interactions of economic actors and the collapse of the wave function. Von Neumann's seemingly divergent interests had a funny habit of colliding with each other in interesting ways.

Morgenstern continued to ingratiate himself with the IAS faculty and with the physicists and mathematicians in particular. His dissatisfaction with Princeton only grew with the comparison. 'There is a spark missing in the department,' he complained. 'It's too provincial.'[37] Worse luck, his manoeuvring would never win him the prize he craved: Morgenstern stayed at Princeton until he retired in 1970. He did, however, successfully make a friend of von Neumann.

Morgenstern was soon obsessively seeking von Neumann's opinions on his work and craved the mathematician's approval. He was pleased that von Neumann shared his disdain for contemporary economics. The discipline needed 'to introduce new thought forms', Morgenstern wrote.[38] As to what these might be, Morgenstern had not a clue. 'The bad thing for me is that I see all this, and feel that it is necessary, but suspect darkly that it escapes me,' he lamented. 'All this probably because I never had the necessary wide mathematical training. Sad.'[39]

The pair soon realized they had common ground. Morgenstern's 'unsolvable' problem of a seemingly endless chain of reasoning was exactly the sort of question that might be addressed by game theory. Could something useful be gleaned about the response of millions of economic actors by assuming they were all out for number one? If so, the circularity of guess and counter-guess might be made to converge on a set of optimal strategies for all players.

The question reignited von Neumann's interest in game theory. He had after all solved the problem for two-person games: in the example of cake-division, a forecast of the outcome would not alter either sibling's choice of strategy. Von Neumann began thinking about extending his theory to encompass games with any number of players. Halperin, who called on von Neumann several times a week during the summer of 1940, found him completely engrossed in his task. Sometimes von Neumann would talk to him non-stop for an hour and a half, says Halperin, but at other times 'he would stand apart, deep in thought, his brown eyes staring into space, his lips moving silently and rapidly, and at such times no one ventured to disturb him'.[40]

Morgenstern too had started a paper introducing economists to game theory. When von Neumann suggested they write it together, he was delighted. 'Have started a treatise with Johnny about games,' wrote Morgenstern on 12 July 1941. 'What fun. We will probably get it finished before September.'[41] He imagined the final product would be a 'pamphlet of about 100 pages'. But what followed, said Morgenstern, was a 'period of the most intensive work I've ever known'. Von Neumann was busily zigzagging the country, answering the call of America's military bigwigs. 'As soon as he got in,' Klári notes, 'he called Oskar and then they would spend the better half of the night writing the book.'[42]

Morgenstern's unhappiness over the state of economics and his weak grip on mathematics grew. 'I neither can nor wish to let go of set theory,' he confided to his diary. 'I was an idiot not to have studied math even as a sideline at the University of Vienna, instead of this silly philosophy, which took so much of my time and of which so little is left.' He was in thrall to von Neumann's genius – and unsettled by it. 'Johnny called me; he likes my manuscript . . . I am very happy about this,' he wrote the same day, adding: 'He is working continuously without a break; it is nearly eerie.'[43]

Weeks lengthened into months. Morgenstern, unable to contribute meaningfully to any of the technical aspects of the burgeoning theory, stimulated economic discussions and served as a foil to von Neumann by asking interesting questions.[44] Tired of seeing the two men locked in seemingly perpetual collaboration, Klári, who collected ornamental elephants, declared she would have nothing more to do with the book unless it featured one. Her pachyderm duly appears under section 8.3 in a set-theoretic diagram, hiding in plain sight.

Even Princeton University Press, cowed by the book's growing length, threatened to pull the plug on the project. In April 1943, however, the 'pamphlet' was finished: 1,200 pages of typescript were finally despatched to the publisher, landing with a thump on the editor's desk.

Morgenstern's principal contribution to the volume was the introduction – which would become the most widely read part of *Theory of Games*. By the time he wrote it, he was thoroughly disillusioned

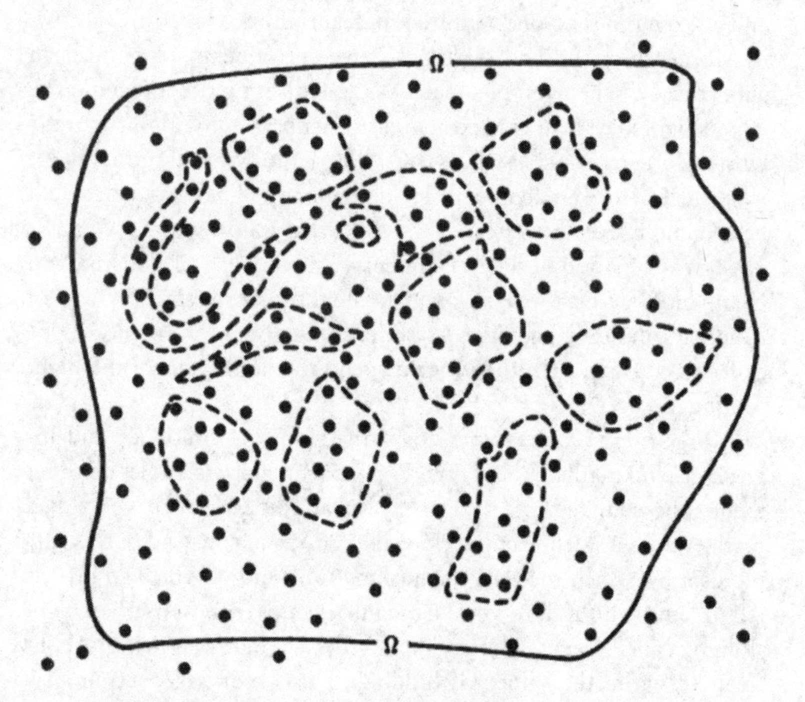

How do you hide an elephant? *From* Theory of Games.

with his discipline. 'Economists simply don't know what science means,' he complained privately. 'I am quite disgusted with all of this garbage.'[45] John Maynard Keynes, whose thinking shaped government policy the world over for much of the twentieth century, was 'one of the biggest charlatans who has ever appeared on the economic scene', said Morgenstern. 'And everybody is on their belly before him.'

Morgenstern strikes a more civil note in his introduction, but the message is much the same: the emperor has no clothes. *Theory of Games* opens with a damning assessment of the economics of the time. The social problems that economists were eager to address had not been formulated precisely enough to be analysed mathematically. The behaviour of individuals had been neglected altogether, even though their decisions in aggregate swayed the economy – just as the motion of molecules determine the bulk properties of a gas.

In part, the lack of meaningful progress in economics was due to a dearth of data. Newtonian mechanics, which revolutionized physics in the seventeenth century, was built on systematic observations made by astronomers over several millennia. 'Nothing of this sort has occurred in economic science,' says *Theory of Games*. Sweeping generalizations had been made on the basis of very little evidence. Mathematics had been used as window dressing to disguise the perilously weak foundations on which the discipline stood. 'Economists frequently point to much larger, more 'burning' questions, and brush everything aside which prevents them from making statements about these,' the book continues. 'The experience of more advanced sciences, for example physics, indicates that this impatience merely delays progress.' The modest goal of *Theory of Games* was to begin addressing these shortcomings by capturing the simplest forms of interaction in rigorous mathematical terms. After this provocative opening followed more than 600 pages of dense explication of proofs drawn from set theory and functional analysis, written mainly by von Neumann.

Theory of Games opens by considering the case of an individual pitted against nature – a 'Robinson Crusoe' economy. With no one else around to get in his way, Crusoe can satisfy any of his desires that his desert island allows. If he adores coconuts, he is in luck. Cravings for filet mignon or Beethoven's Fifth will probably have to go unsatisfied. This is an easy mathematical problem to solve: a matter of fulfilling Crusoe's desires to whatever extent his island allows. In game theory, the situation is akin to a single-player game like patience. Crusoe's choice of strategy will be determined by whatever resources are available to him on the island, just as a patience player's most effective strategy will depend on the order of cards in the deck.

What if another person, Friday, arrives on Crusoe's island? Then the problem of obtaining some optimal outcome is very different. Unfortunately, this is no longer a simple matter of maximizing each person's desires within the constraints of the island's resources. Once any overlap at all between their needs exists, there is immediately a conflict of interests. 'This is certainly no maximum problem but a peculiar and disconcerting mixture of several conflicting maximum problems,' von Neumann and Morgenstern note.

The tools of classical mathematics, such as calculus, are of no use in this situation, they explain, shocking many economists, who were only just catching up with the use of calculus in their field. 'A particularly striking expression of the popular misunderstanding about this pseudo-maximum problem,' the pair continue, 'is the famous statement according to which the purpose of social effort is the 'greatest possible good for the greatest possible number.' A guiding principle cannot be formulated by the requirement of maximizing two (or more) functions at once.'

The correct approach, of course, is that of game theory. Von Neumann explains that he will first thoroughly explore two-player zero-sum games, then generalize the theory to cover games with any number of players in which the payoffs do not necessarily sum to zero.

First, however, any theory that claims to be able to identify a person's 'best' strategy needs a simple way to quantify their likes and dislikes. An entrepreneur's goals can be easily measured in terms of costs and profit. But there is more to life than money, and the arena of game theory includes any situation where conflict can arise.

The orthodoxy in economics was that preferences could not be measured and put on a numerical scale but only ranked. When Morgenstern told him this, von Neumann quickly invented a revolutionary theory that allowed an individual's likes and dislikes to be assigned a number on a 'happiness' or utility scale, just as a thermometer reading gives the temperature of a bowl of soup. 'I recall vividly,' says

Morgenstern, 'how Johnny rose from our table when we had set down the axioms and called out in astonishment: '*Ja hat denn das niemand gesehen?*' ('But didn't anyone see that?')[46] Apparently, no one had.

First, von Neumann says, we have to accept 'the picture of an individual whose system of preferences is all-embracing and complete'; that is, given a choice between any two events or objects, the person in question is always able to choose the one which they prefer (e.g. would you prefer to go to the cinema or to watch TV and get a takeaway?). Furthermore, let's say that not only can they choose decisively between two different options, they can choose between combinations of events, e.g. we've left it a little late to go to the cinema, so there's a 50 per cent chance we might miss the film, in which case we will definitely go bowling instead. Or we could just get takeaway . . .

As long as these modest assumptions hold, von Neumann says, one can calculate utility scores (the conventional units are *utils*). To calibrate a utility scale, pick a pair of events.[47] One must be your most feared calamity – assign this a utility score of 0 utils. The other is the most marvellous experience you can realistically imagine – give this a score of 100 utils. This is akin to using the freezing and boiling points of water to calibrate a Celsius thermometer.[48]

Imagine now that it is your birthday, and your favourite person has organized a wonderful party with lots of presents, jelly and ice cream and a huge chocolate cake. Hey presto! John von Neumann appears in a puff of smoke. He is in one of his devilish moods and has a proposition for you. Instead of your birthday party, you can have a lottery ticket that will give you a chance of winning your 100-util vision of heaven. The downside is if you lose, you will instantly be condemned to living out your nightmare. The question is what would the probability of winning have to be before you are willing to accept von Neumann's offer? If the ticket would need to give you a 75 per cent chance of winning then your birthday party is worth $\times 100 = 75$ utils. A more risk-averse person might need a ticket with a 98 per cent chance of winning the big prize before they would trade. The party would therefore be worth 98 utils to them.

Von Neumann can now define what 'rational' means in the context of the theory. He proposes that a player is behaving rationally if he adopts a strategy in any situation that ultimately serves to maximize his gains against other rational players. This payoff, von Neumann explains, 'is, of course, presumed to be a minimum; he may get more if the others make mistakes (behave irrationally)'. A description of 'rational behaviour' then boils down to 'a complete set of rules' that tells a participant how to play in every situation that may arise in the game to achieve this aim.[49] This simplifies the mathematics enormously because, thanks to utility theory, everything a player strives for is summarized by a single number.

Von Neumann had achieved the supposedly impossible – a rigorous way to assign numbers to nebulous human desires and predilections. 'To this day the most important theory in the social sciences' was how Nobel laureate Daniel Kahneman described von Neumann's accomplishment in 2011, more than sixty years after *Theory of Games* first appeared.[50] The influence of utility theory and the notion of the rational calculating individual that is at its heart would quickly reach far beyond the ivory tower.

Armed with utility theory, von Neumann begins his analysis of two-player games. Some such games are simple enough to allow all rational moves to be specified completely. The best strategy for noughts and crosses (tic-tac-toe) is easy enough to work out – the rules of optimal play can be written down on a single sheet of paper. As most kids quickly realize, if both players make only optimal moves, the game inevitably ends in a draw.

Von Neumann gives two ways to represent games which are still used by game theorists today: the 'extensive' form and the 'normalized' form. Both are equivalent, and von Neumann advises using whichever is most convenient for the problem at hand.

The extensive form resembles a tree. Each move is a branch point called a node. The branches radiating out from each node are possible moves. At the end of each branch is a leaf, corresponding to the final outcome of the game. If someone is bored enough on a wet afternoon, they might be able to draw the complete diagram for tic-tac-toe. The shortest branches will be five moves long – the minimum number of moves the first player needs to win. The longest can only be nine nodes long: the game must end at this juncture because the grid is full.

The tree for chess, however, gets very cluttered very quickly. After

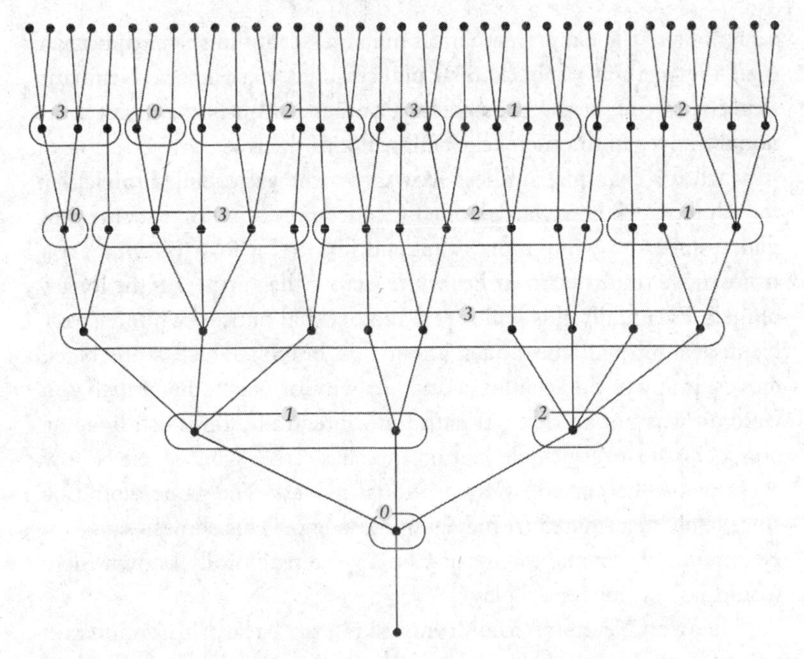

Von Neumann and Morgenstern's illustration of the extensive form of a game.

three moves, the pieces on the board can be in one of 121 million different possible configurations. Between helping to invent the design principles of digital circuits and modern information theory, American mathematician Claude Shannon calculated that there are at least 10^{120} possible games of chess – comfortably more than the number of elementary particles in the universe.

Von Neumann called chess and tic-tac-toe games of 'perfect information' – all the moves in the game are visible to both players. He showed that all two-player, zero-sum games of perfect information have a 'solution' – as long as the game does not go on for ever.[51] Tic-tac-toe obviously fulfils the criterion. Less obviously, chess does too, because there are various rules that end a game in a tie: if neither player has enough pieces left on the board to checkmate, for example, or if the same position occurs three times.

All games of perfect information must end in a win or a draw, and, crucially, there is always only one optimal move for each player at each node of the game tree. Von Neumann's proof of this proposition used the method of 'backward induction', a mathematical process akin to starting at the ends of each branch and pruning away 'irrational' moves until one reaches the trunk of the tree.

Start, for example, at a leaf that shows the game ended in victory for white. Work backwards node by node and cut away any branches that result from white moves that end in a tie or loss for white. If a white move results in an earlier white victory then prune off the longer branch. Eventually this will reveal the optimal path to white victory. Next, examine all the nodes along this branch which were black moves. If any of these end in a draw or win for black, the branch you were on was not an optimal path for white at all: lop it off. Imagine doing this for every single leaf on the chess tree. What is left is how chess would be played by two rational players. The game would be completely determined from beginning to end. 'This shows,' says von Neumann, 'that if the theory of Chess were really fully known there would be nothing left to play.'

When von Neumann told Bronowski in the back of a London taxi that 'chess is not a game' he meant exactly this: that a computer could be programmed to find this 'perfect' game. Chess aficionados can rest easy, though. No machine on Earth is close to powerful enough to carry out the task. No one even knows if a completely rational game of chess ends in a win for white or a draw. There is a small – but finite – chance, however, that someone, somewhere, perhaps sitting in their living room wearing a dressing gown, has already played it.

The extensive form of a game can get messy. The normalized form (now also known as the 'strategic' or 'normal' form) presents a game's payoffs (utility scores) as a table. Von Neumann gives several examples. The normalized form of Matching Pennies looks like this:

	Heads	Tails
Heads	1 penny	-1 penny
Tails	-1 penny	1 penny

The normalized form of Matching Pennies, with payoffs shown for the player who wins when both pennies match.

Other games which at first glance look nothing like Matching Pennies may on closer inspection be two-player zero-sum games. To demonstrate this, von Neumann solves Morgenstern's 'impossible' problem from Conan Doyle. He finds optimal strategies for Holmes and Moriarty, and the chance of the detective making good his escape.

Von Neumann assigns Moriarty 100 utils for catching and killing Holmes at either Dover or Canterbury. Holmes making good his escape to the Continent results in a loss of 50 utils for Moriarty. Should Holmes evade Moriarty by getting off at Canterbury because Moriarty took the fast train to Dover, it is a tie, because Holmes has failed to reach the Continent, and the chase goes on.

Holmes / Moriarty	Dover	Canterbury
Dover	100	0
Canterbury	-50	100

Moriarty's payoff table.

Going to Canterbury carries the risk of incurring the worst payoff of -50, because Holmes gets clean away. On any particular day, Moriarty will prefer to go to Dover. If he favours Dover too predictably, however, Holmes will out-guess him and get off the train at Canterbury. So Moriarty must keep Canterbury in the mix of options.

Moriarty's optimal strategy is to take the fast train to Dover with a probability of 60 per cent. He should get off at Canterbury with only a 40 per cent probability. That way, Moriarty's average payoff is the same (40 utils), whatever Holmes chooses to do.[52] Holmes's strategy is the reverse: get out at Canterbury with a likelihood of 60 per cent, otherwise stay on the train until Dover.

In 'The Final Problem', Holmes alights at Canterbury, hiding behind luggage as Moriarty roars past on a special train to Dover. Both had taken the most probable course described in von Neumann's analysis, but the odds were actually stacked in Moriarty's favour. With von Neumann's assumptions, 'Sherlock Holmes is as good as 48% dead when his train pulls out from Victoria Station' – and he has only a 16 per cent chance of making a clean getaway. With his supposedly 'insoluble' problem neatly solved, a chastened Morgenstern recants his earlier pessimism in a footnote.

Von Neumann had promised that a game-theoretic analysis of poker would follow his 1928 paper. Now, some fifteen or so years later, he provides one. Poker is a game of imperfect information. In many ways, it is *the* game of imperfect information. That a player does not know what cards other participants are holding is what makes the game fun and exciting at all. An optimal strategy for poker must embrace this uncertainty. The key to winning, as anyone who has played the game knows, is the art (and science) of bluffing: betting high with bad hands.

'It was Von Neumann's analysis of Poker that made me into a game theorist,' says Ken Binmore, who helped design auctions used to sell off unused radio bandwidth to telecoms companies in Britain.

> I knew that good Poker players bluff a lot, but I just didn't believe that it could be optimal to bluff as much as Von Neumann claimed. I should have known better than to doubt the master! After much painful calculation, I not only had to admit that he was right, but found myself hopelessly hooked on game theory ever after.[53]

Von Neumann considers a two-player poker game to keep things simple. A third of the way into his book, he has only discussed two-player zero-sum games. There are 2,598,960 combinations of the five-card hand each player starts out with, so von Neumann abstracts away that complication by saying each player is given a number instead – let's say between 0 and 100 (1–99). All numbers are equally likely to be dealt, so if one player is dealt a 66 then she will correctly assume that she is about two times more likely to have a higher card than her opponent.[54]

Poker comes in many flavours. In 'draw' variants, for example, a player can exchange some of the cards in his hand for others, and there are several rounds of bidding. Von Neumann dispenses with these complexities too. In his toy model, there will only be two bids allowed: high (£*H*, i.e. for an amount *H* pounds) or low (£*L*), and only one round of betting. The game is 'risky' if the 'high' bid is much greater than the 'low' bid.

The two players make their initial bids simultaneously (again deviating from real poker, where one player goes first), then compare their bids. If both players have bid high or low, they compare 'cards', and the player with the stronger 'hand' receives £*H* or £*L*. If both players were dealt the same number, then they keep their bets. If their bids are different, then the player with the low bid has a choice. She can 'pass', only paying the other player £*L*, and neither player's hands are revealed. Or she can 'see' by changing her bid to 'high', at which point the players compare hands as if they had both bid 'high' in the first place.

Those are all the rules there are to von Neumann's version of poker, and they are sufficient to explain the logic of bluffing. He determines the minimax strategy for both players is, naturally, to bid high with a good hand. The threshold beyond which they should always bid high is determined by the relative sizes of the high and low stakes. If the high bet is twice the low, then anything above 50 warrants the bigger stake; if it is three times higher, then above 66; four times, and betting high is only recommended for a card above 75. The pattern is clear: better cards are required in 'riskier' games (i.e. those in which the high bid is much greater than the low) to bid high.[55]

Von Neumann then considers what players should do if they draw a card with a value below the threshold. In these cases, he finds that while the players should bid low most of the time, they should throw in an occasional high bid. How often players should bid high with a weak hand is again dependent on the ratio of high to low stakes. For 2, 3 and 4 times, players should bid high with probability a third, a quarter and a fifth, respectively.[56] These are nothing other than the bluffs of real poker.

'Due to the extreme simplifications which we applied to Poker for the purpose of this discussion, "Bluffing" comes up in a very

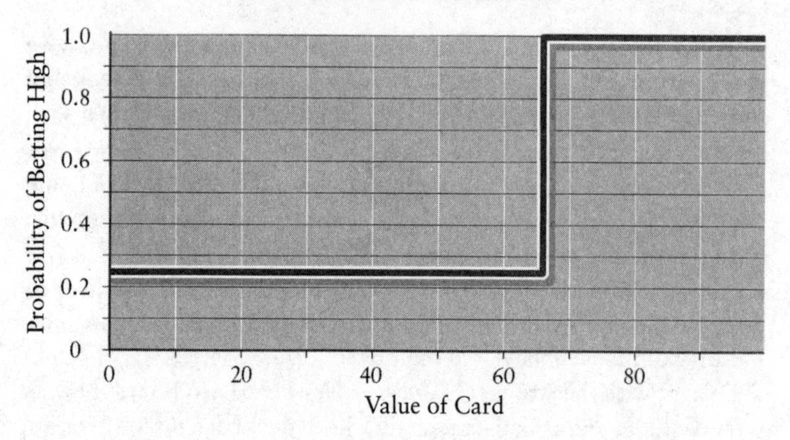

Optimal Strategy for von Neumann's poker. High stake is three times value of low stake in this example.

rudimentary form only; but the symptoms are nevertheless unmistakable,' von Neumann says.

Bluffing, however, may serve one of two purposes. A high bid with a weak hand may fool an opponent into believing the player holds a strong hand, inducing her to pass. Or a bluff's primary objective could be to keep an opponent guessing whether a player's high bid necessarily means she holds a strong hand. If someone only bids high with good cards, their opponent will quickly learn to pass – reducing their winnings.

Von Neumann demonstrates that the latter is the better explanation: a failure to bluff does not lead to heavy losses if the opposing player sticks to the optimal strategy (the hands they would have won with bad cards are balanced by the hands they would have lost when their opponent calls their bluff). However, a smart opponent will quickly depart from their 'optimal' strategy by passing whenever the non-bluffing player bids high and throwing in more bluffs themselves. As Binmore says, 'The point of bluffing is not so much that you might win with a bad hand, as that you want to encourage the opposition to bet with middle-range hands when you have a good hand.'[57]

*

Poker rounds off von Neumann's exhaustive treatment of zero-sum two-person games. He had laid down the axioms of game theory and given formal definitions of terms such as 'game' and 'strategy'. He also offered a proof of his minimax theorem, which was much more elementary than the original of 1928. In 1938, Jean Ville, a student of Émile Borel, had published a simple algebraic proof of minimax. After Morgenstern chanced upon Ville's work in the IAS library, he alerted von Neumann, who made Ville's proof simpler still. For the remainder of *Theory of Games*, von Neumann tries to extend his results to games with any number of players and to situations where there is the possibility of mutual benefit. Here he flounders. Much of what was to come in game theory after von Neumann and Morgenstern's monumental book would be forged by those rushing to fill the void.

The approach to multi-player and non-zero sum games set out in *Theory of Games* entails finding tricks to reduce them to the zero-sum two-person case. Since von Neumann's minimax theorem guarantees that every such game has a rational solution, he thought that analysing these 'fake' two-player games would reveal the best strategy for each side.

Von Neumann starts by considering a three-player game of strategy. He reasons that, rather than go it alone, the players would try to team up to edge out one of their opponents. Anyone who has played a three-cornered game of Monopoly or Settlers of Catan will have noticed such scheming occurs quite naturally, with barely a word needing to be said. 'Every man for himself' quickly becomes a joint effort by the two weaker players to prevent the player who has taken an early lead from winning outright. How early in the game this unholy alliance forms and how long it lasts helps to decide who wins and who loses.

Monopoly or Catan are far too complex to be useful models for a three-player game. Von Neumann instead invents a simple 'game' in which the only object is for two players to form a 'couple', leaving the third player out in the cold. Each player secretly writes down the name of one other player, and then all three reveal their choices. If two players have chosen each other, they get half a point each, and the third loses a point. Otherwise no one gets anything. A rational player quickly grasps that the only way to win is to negotiate 'off-stage' before the match starts and strike a deal. Even this pointless-seeming game raises interesting questions that also crop up in everyday life. For instance, how does anyone know that a promise will be kept? If the promise is actually more like a contract, then what mechanisms are necessary to enforce it? 'It is quite instructive,' says von Neumann, 'to observe how the rules of the game are absolutely fair, but the conduct of the players will necessarily not be.' In fact, he says, the negotiations themselves could be treated as a game – which is what some later theorists did.

In *Theory of Games*, von Neumann adds complexities to his model game. Instead of an even split between the players successfully forming a coalition, one player of the three insists on getting more; and anyone partnering with her gets less. This 'greedy' player will, of course, never be able to form a coalition unless she agrees to forgo her extra slice of the pie; the other two will always pair off to maximize their payouts.

Likewise, if two players now get some extra fraction of the payout in a coalition with the third, they will lose interest in pairing up with each other and instead start bidding for player three's cooperation. Logically, a 'couple' will form only when one of the two supposedly advantaged players compensates player three fully. With a system of compensation (bribes?) and bidding in place, von Neumann is able to determine the mathematical conditions under which players form coalitions or all decide to play solo. In the latter case, each player gets a payoff dependent on the strategies chosen by themselves and the two others. If, on the other hand, a coalition forms, von Neumann showed that the situation can be treated as a two-person zero-sum game, with the accomplices treated as a single 'player' versus the remaining participant.

So far, so good. There are, however, some cracks appearing in the analysis. With three players, the theory identifies three possible coalitions.[58] Von Neumann has limited his theory to a 'static' one-off game. Otherwise, there would be nothing to prevent a bloody *Game-of-Thrones*-style cycle of alliance and treachery that continued indefinitely.[59] Still, despite this constraint, the theory has little to say about which of the three possible coalitions will *actually* form. Von Neumann is forced to acknowledge that in many circumstances there may not be a stable, unique solution to a game at all. Hardly the complete description of rational play that the early pages of *Theory of Games* had promised. Von Neumann and Morgenstern try to mitigate this by arguing that in the 'real world' a particular coalition might be stable because of what they referred to as the 'established order of society' – the dominant cultural norms of a particular time or place. If 'discrimination' is tolerated, even rampantly unfair solutions are permissible, they say. At one extreme, for instance, if negotiating with one player is taboo, then only one coalition can form. With these caveats, von Neumann declares the zero-sum three-person game solved.

The complexity of the four-person game means von Neumann is forced to concentrate on a number of special cases. He finds situations that can be reduced to the two- and three-person games he has solved already. Games between two pairs of players or a triplet versus a (losing) singleton, for instance, can be treated as if they were two-person zero-sum games. A 'core' coalition of two players trying to attract a third to form a winning team can be modelled as a three-player game. Von Neumann's analysis of five-person games is even more limited. He is only able to discuss symmetric games: those where the payoff for a strategy depends only on the other strategies deployed and not who plays them. Symmetric games are important: they reflect social situations where everyone wants the same thing. (Tensions over water, for example, have led to international conflict – and cooperation.)[60] But von Neumann can only describe two instances of the symmetric five-person game, where the balance of payoffs means that, to win, a player has to form a coalition with either one or two others.

At last, von Neumann is able to begin his attack on *n*-person games – those with any number of players. 'It is absolutely vital to get some insight into the conditions which prevail for the greater values of *n*,' he says, as 'these are most important for the hoped for economic and sociological applications.' He concedes, however, that 'In the present state of things we cannot hope for anything systematic or exhaustive.'

Von Neumann first looks at a subset of simple games which can be neatly split into separate, smaller games. Such 'decomposable games' are each played by a 'self contained group of players, who neither influence, nor are influenced by, the others as far as the rules of the game are concerned'. He describes how to recognize games that can be broken down in this fashion and some of the coalitions that might form under these circumstances.

Next, he tries to find a more general solution to the problem. He defines what game theorists today call the 'stable set': a group of solutions that cannot be trumped by any others. Each solution is a set of 'side-payments' made between coalition partners[61] to ensure their loyalty. The coalition is stable because none of its members can cut a better deal. Von Neumann is able to show that some three-person games have solutions in the form of stable sets. As before, however, there are a multitude of possibilities with no mathematical criteria for choosing which one of these will actually be realized. Which solution ultimately wins is determined by prevailing 'standards of behaviour'.

So far, von Neumann has dealt only with games that are zero-sum. But life is rarely a game of total conflict. Economic growth is not zero-sum: the world is a more prosperous place now than it was 200 years ago. Often a situation really is win-win. Sometimes, everyone loses. 'Zero-sum games are to the theory of games what the twelve-bar blues is to jazz: a polar case, and a historical point of departure,' says economist Michael Bacharach.[62]

Von Neumann and Morgenstern knew they had to offer some way of tackling non-zero-sum games if their theory was to say anything useful about 'economic behaviour', as their book's title promised. Towards the very end of *Theory of Games*, they do. The approach is something of a mathematical sleight-of-hand, lacking the rigour of von Neumann's earlier treatment of two- and three-player zero-sum games. He introduces a passive 'fictitious player' whose only role in a game is to act as a kind of utility bank, losing the sum of whatever the other players win or winning what they lose. An n-person non-zero-sum game is transformed with the addition of this phantom into a zero-sum affair played by $n+1$ participants, which can be solved by applying the machinery that has already been developed over hundreds of pages. Von Neumann is at last ready to bring game theory to bear on the simplest of economies.

Contemporary models of the economy were rooted in the work of the French economist Léon Walras. Considered the father of general equilibrium theory, Walras developed his equations in the 1870s with the assumption of perfect competition. Under these idealized conditions, buyers and sellers are so numerous that no individual can by themselves affect the price of goods. The economy reaches equilibrium after prices fall or rise until supply meets demand or vice versa. The formation of a monopoly or oligopoly (a few sellers) is in Walrasian theory a temporary aberration that is eliminated in the long run by the magic of the market.[63]

By the time von Neumann and Morgenstern embarked on their opus in the 1940s, economists had begun to recognize that monopolistic competition was not the exception but the rule. In the mid-twentieth century, the Big Three or Four were the behemoths of the oil and car industries – Standard Oil, or Ford and General Motors. Today the monopolists are more likely to be drawn from the ranks of tech giants such as Facebook, Apple, Amazon or Google.

Theory of Games illustrates why, without tough anti-trust laws and incessant vigilance, monopolies and oligopolies spring up like weeds. In a market dominated by one or a few big firms, each would naturally use its heft to maximize profits. Even with no active collusion, they push up prices for consumers as if they had formed one of von Neumann and Morgenstern's coalitions. No additional assumptions were required to reproduce the phenomenon – von Neumann's original axioms sufficed.

The actual discussion of the economic applications of game theory in the book itself is brief and revolves around the results for non-zero-sum games of one, two and three players. For the one-person 'Robinson Crusoe' case, the result is, as expected, a simple maximization of an individual's gains, limited only by the available resources – a centrally planned communist economy. The two-player market, with a buyer and a seller, is known as a bilateral monopoly. The solution is in line with common sense. The agreed-on price for whatever is being sold will be somewhere between the maximum the buyer is willing to pay and the minimum price at which the seller will part with it. Where exactly in this interval the two parties agree to settle depends on 'negotiating, bargaining, higgling, contracting and recontracting' before the transaction. On this matter, the *Theory of Games* is silent.

There are too many possibilities for a comprehensive discussion of three-player markets, so von Neumann focuses on a scenario in which there are two buyers seeking to secure some indivisible commodity from a single seller. The most straightforward outcome is that the stronger buyer outbids his competitor, buying the goods for a price somewhat above whatever his competitor is willing or able to pay. However, von Neumann identifies another more interesting solution in which the two buyers form a coalition. Now together they can haggle with the seller, depressing the price to below the weaker buyer's maximum price and perhaps as low as the absolute minimum the seller is willing to take. The pair can then divide up between themselves whatever the stronger buyer has saved.

Von Neumann does not discuss in detail the related situation of two sellers and one buyer (a 'monopsony'). Rather cheekily, he leaves the maths as an exercise for the reader. But by analogy with the case of two buyers, either the sellers compete and the buyer picks up the product cheaply (below the more expensive seller's minimum price) or the sellers form a coalition, driving up the price until they hit the unfortunate buyer's spending limit.

The discussion of the one-, two- and three-player market games is more or less the sum total of the offerings to economists in *Theory of Games* – no more than a tantalizing hint of its potential. Luckily, game theory's applications to the business world soon had an eloquent, if unlikely, champion in the form of the American business journalist John McDonald. An ardent Trotskyite, McDonald travelled to Mexico in 1937 to serve on his hero's secretarial staff. In 1945, he joined *Fortune* magazine and wrote a string of books and articles popularizing game theory with the help of *Theory of Games*' two authors.

Perhaps because of his experience chronicling the clashes between the industry titans of capitalist America, McDonald appreciated how well the theory could model real economic interactions. 'The importance of the three-person coalition game for economics is its dissection of the phenomenon of "oligopoly" and "monopolistic competition," which has long baffled economic thought,' he wrote in 1950. 'Unlike any others, von Neumann and Morgenstern build coalitions integrally into their theory.'[64]

McDonald describes two grocers being undercut by a supermarket in game-theoretic terms:

> 'The two grocers make a coalition against the consumer; they play a two-man game with each other which results in taking more money from the consumer. Along comes the supermarket offering prizes or payments to the consumer in the form of lower prices (based on higher output and lower costs). Supermarket and consumer make a coalition . . . against the grocer coalition. The supermarket receives payments in profits (from larger volume at lower prices for each consumer); the consumer receives payments in savings. But the game is not over when the first grocer coalition retires from the field. If the supermarket has other competitors, as it usually does, the consumer can maintain a strong position by threatening coalitions with those competitors. But if the supermarket and its competitors now combine, they could make additional gains through higher prices for a while at the consumer's expense. The situation with which the game began would then be restored, to be upset, perhaps, by another newcomer.'[65]

Von Neumann and Morgenstern's *Theory of Games and Economic Behavior* appeared in 1944. The first edition promptly sold out – an article about the book on the front page of the *New York Times* and a dozen glowing reviews in prestigious journals turned the opus into an unlikely best-seller. 'Ten more such books,' declared one reviewer, 'and the progress of economics is assured.'[66]

Despite the many plaudits, game theory did not immediately catch on with economists. The book was too mathematical. Even at Princeton, where the mathematics department quickly became a hotbed for research on the subject, the economics department was hostile. Personalities played a role. Von Neumann was much admired but, as a non-economist, he was an outsider. Morgenstern's supercilious manner rubbed many people up the wrong way. 'The economics department just hated Oskar,' says economist Martin Shubik, who arrived at Princeton in 1949 expressly to study game theory, 'not nearly just alone because they couldn't understand what was going on, but there was a certain aristocratic touch to Oskar, and . . . that was an extra reason for the hate.'[67] Economist Paul Samuelson, a future Nobel laureate, was watching from his berth at MIT. Morgenstern was 'very Napoleonic', he later told the historian Robert Leonard, given to making 'great claims' that he did not have the wherewithal to prove.[68]

The main reason for the disdain, though, was that game theory had not yet proved its worth for dealing with economic questions. There were too many loose ends. Top of the list was von Neumann and Morgenstern's 'cooperative' solutions to games with more than two players. *Theory of Games* assumes that utility can be transferred seamlessly between players in a coalition. This was clearly not true if the 'gains' in question are not bundles of cash but troublesome even if they are – a ten-pound note is worth more utils to someone who is homeless than to a millionaire. Moreover, *Theory of Games* provides no method for calculating how much each player in a coalition should receive: what was a 'fair' settlement?

Second, von Neumann and Morgenstern's 'stable-set' solution for the *n*-person game proved contentious. Did *all* multi-player games have a cluster of coalitions that could not be disrupted by a member finding a better deal? Von Neumann did not prove it. A quarter of a century after the publication of *Theory of Games*, mathematician William Lucas would find a ten-person game which had no stable sets at all.[69]

Many also balked at von Neumann's approach to non-zero-sum games. The use of a fictitious player, notes mathematician Gerald Thompson, 'helped but did not suffice for a completely adequate treatment of the non-zero-sum case. This is unfortunate,' he adds, 'because such games are the most likely to be found useful in practice.'[70]

Von Neumann's biggest blind spot proved to be his failure to consider games in which coalitions were either forbidden or players could not, or simply did not want to, team up. As game theory gained a reputation for a relentless focus on cut-throat competition between calculating individuals, its progenitor did too. Yet the idea that anyone would choose to go it alone when cooperation produced better results was foreign to von Neumann's central-European temperament. He did not think that was the way the world worked. 'To von Neumann,' says Leonard, 'the formation of alliances and coalitions was *sine qua non* in any theory of social organisation.'

Theory of Games was not the complete guide to strategic behaviour that von Neumann and Morgenstern hoped. The two-person zero-sum game, perhaps the most elegant and immediately applicable part of the book, was rooted in von Neumann's minimax theorem, first developed nearly twenty years earlier. Non-zero-sum games with an arbitrary number of players were a work in progress. However, von Neumann's revolutionary work on utility, together with the rigorous description of games and their representations in extensive and normal form, would be the bedrock on which other talented mathematicians would soon build. A theory that sparkled with von Neumann's stardust would prove irresistible to some of the keenest minds of a new generation such as John Nash, Lloyd Shapley, David Gale and others. By the 1960s, their work would open the floodgates in economics and the wider social sciences.

In 1994, fifty years after the publication of von Neumann and Morgenstern's *Theory of Games and Economic Behavior*, the committee tasked with choosing the winner of the Nobel Prize in the economic sciences were faced with a tricky decision.[71] They agreed that the award should go to a game theorist to mark that canonical book's anniversary, but with both its authors long dead, they were split over who should win. At the last minute, Nash who had recently recovered from decades of mental illness, was awarded the prize with two other game theorists, Reinhard Selten and John Harsanyi.

Game theory quickly proved itself worthy of the accolade in spectacular style. The same year the three laureates would receive their gold medals from the king of Sweden, the US government was preparing to auction bands of the radio spectrum to telecoms firms. Thousands of licences worth billions of dollars were at stake. Many past sell-offs had flopped.[72] In New Zealand, a botched 'second-price' auction, in which the winner only pays the second-highest bid, resulted in a firm that bid NZ$7 million paying NZ$5,000; and a university student picking up a licence to run a television network for a small city for nothing – because no one else had bid.

Under pressure to avoid such pitfalls, the Federal Communications Commission (FCC) asked for proposals and adopted a system designed by game theorists including Paul Milgrom and Robert Wilson. Under the rules of the 'simultaneous ascending auction' they devised, several licences go up for sale at once, and bidders can bid for any they like.[73] The whole, painstakingly designed edifice rested on the non-cooperative game theory advanced by the three Nobel laureates of that year. And it was a huge success.

Ten licences for paging services were sold by auction in July and raised $617 million. The following year, a new round of auctions for ninety-nine communications licences, hailed by the *New York Times* as 'The Greatest Auction Ever', generated $7 billion. By the end of 1997, insubstantial airwaves had netted a very tangible $20 billion for the US government – more than twice their estimated worth.[74] Other countries soon followed the American example. In 2020, just over a quarter of a century after the first US spectrum auctions took place, Milgrom and Wilson would share the Nobel prize themselves.

A deluge of Nobels would go to game theorists after Nash, Selten and Harsanyi. Thomas Schelling and Robert Aumann picked up the prize in 2005 for their work on conflict and cooperation. In 2012, the eighty-nine-year-old Shapley, who advanced cooperative game theory, got the call from the Nobel committee – and told them he 'never, never in my life took a course in economics'. His co-winner, Alvin Roth, had used the matchmaking algorithm Shapley helped design to pair junior doctors with hospitals, students with schools – and kidney donors with patients.

In 2009, Elinor Ostrom became the first woman to win the economics Nobel.[75] She had travelled far and wide to apply game theoretic analyses to the governance of the 'commons' – resources used by many people that may be depleted by their collective actions.[76] Ostrom described how locals invented ways to protect such resources. In Nepal, for instance, she found that cattle belonging to farmers who had failed to follow rules for water usage were interned in a 'cow jail' until a fine was paid to secure their release.[77]

Questioning some of game theory's fundamental precepts, as Ostrom did, has produced rich insights. Another economics Nobel laureate, psychologist Daniel Kahneman, challenged game theory's assumption that humans are entirely rational and had preferences and tastes that never changed. An admirer of von Neumann, 'one of the giant intellectual figures of the twentieth century', Kahneman and his close collaborator, Amos Tversky, studied how real people actually make decisions and devised their own 'prospect theory' to explain findings that ran counter to some of utility theory's predictions.[78]

Jean Tirole, the 2014 Nobel winner, used game theory to analyse industries dominated by a few powerful companies – an increasingly pertinent topic in the Internet economy. While economists had well-developed theories for markets with lots of competing firms and for monopolies, the theory underpinning how oligopolies work was sketchy. Tirole provided a framework for understanding – and regulating – these sorts of markets: think Google, with its dominant share of the Internet-search advertising market, or Amazon, responsible for nearly half of all sales made on the Internet in America.

Tech firms themselves have employed leading game theorists to help design online advertising marketplaces, bidding systems, product-ranking algorithms – and ways to stay ahead of regulators.[79] The most useful – and profitable – area of application has been in the realm of auction design – specifically those that determine the price of keywords used to place ads in search results.[80] Keyword auctions are now responsible for a large chunk of income for a swathe of Internet companies – making billions for Google, as well as other firms better known for selling goods rather than ads including Amazon, Apple and Alibaba. Game theorists have since been drawn into every corner of Internet commerce, from pricing cloud computing services to taxi rides, and the design of addictive reward and ratings systems that keep users coming back for more.

The work of Shapley, Nash and others propelled the field into surprising areas. Perhaps the most unexpected area of application was to animal behaviour, where game theory helped biologists to understand how cooperation might evolve in nature, famously 'red in tooth and claw'. One trailblazer was English biologist William D. Hamilton, who first came across von Neumann and Morgenstern's *Theory of Games and Economic Behavior* as an undergraduate at the University of Cambridge in the late 1950s. 'The idea of . . . a biological version of von Neumann's game theory had crossed my mind as soon as I read his earliest account,' he wrote later.[81] Hamilton would, among other contributions to the field, produce a mathematical model of altruism based on the degree of relatedness (kinship) between different organisms. He showed that genes for self-sacrificing behaviours would spread as long as they benefited blood relatives (who were also likely to carry the same genes). Hamilton's theory, now known as 'inclusive fitness', was popularized by Richard Dawkins in *The Selfish Gene*.

Others took Hamilton's work forward, including George Price, a chemist who worked on the Manhattan Project before moving to Britain and making some decisive contributions to evolutionary theory.[82] First, Price wrote an equation that extended Hamilton's ideas to include all evolutionary change – not just traits that benefit family members. This elegant mathematical interpretation of natural selection is now known as the Price Equation. Next, Price worked with the British aeronautical engineer-turned-biologist John Maynard Smith, accounting for many behaviours observed in the natural world by reframing them as 'games' or contests between members of a population of animals. They introduced the concept of an 'evolutionarily stable strategy', which explains why different proportions of animals in a population might have different characteristics.

They illustrated this with the Hawk-Dove game: 'hawks' are unfailingly aggressive, and 'doves' never fight. A population solely composed of doves will quickly be driven to extinction by the appearance of a single hawk in its midst, as the hawk monopolizes food and resources by driving away the peaceable doves. On the other hand, a population comprised only of belligerent hawks would be susceptible to the appearance of a dove – as the dove never fights, it is never injured, so stays in a better condition to find resources. The upshot is that a certain mix of the two strategies – some percentage of hawks and some of doves – is stable, 'fixed' in place by natural selection. A striking real-world example is *Onthophagus Taurus*, a species of dung beetle in which there are 'major' males, which have large horns and fight with each other for access to females, and 'minor' males, which are weedy and hornless and can sometimes sneak past the fighting larger males to breed with females.

Price became so appalled by the idea that altruistic behaviour could be adequately explained by selfishness, rather than by the existence of some nobler motivation, that he took to performing random acts of kindness in a bid to convince himself he was mistaken. He ultimately became so depressed that, in 1975, he ended his own life by slitting his carotid artery with a pair of nail scissors.

Price, Hamilton and Maynard Smith had shown how altruistic behaviours could evolve and become 'fixed' in populations if their beneficiaries were related. But could cooperation ever thrive between unrelated individuals? Computers, biology and game theory came together to show that they could.

Hamilton moved to the University of Michigan from Imperial College London in 1978. A couple of years later, he and Michigan professor of political science Robert Axelrod began to invite academics – including a number of game theorists – to submit strategies for a computer game. In each round, one strategy was pitted against another in a series of 200 prisoner's dilemmas. The logical outcome of a prisoner's dilemma is a bad deal for everyone. In the context of evolution, this might on the face of it suggest that, in interactions with others, animals should behave completely selfishly to short-sightedly maximize their own gains – even when cooperating would improve everyone's lot, including their own, in the longer term. But in real life this does not always happen, and Axelrod realized that a one-off prisoner's dilemma did not reflect the fact that the same two animals might meet many times. Hence the tournament: would the most selfish strategies come out on top? They did not. The winning strategy, called 'Tit-for-Tat', came from game theorist Anatol Rapoport. It was incredibly simple: cooperating by default but behaving selfishly when the opponent did. Axelrod's game had illustrated that cooperation might develop even if animals were evolutionarily inclined to act purely in their own self-interest.

Axelrod and Hamilton's work spawned what Dawkins has called 'a whole new research industry'. Hundreds of papers followed. The mathematics of evolutionary game theory, in which the 'rational players' of von Neumann's game theory are replaced by evolutionary strategies and natural selection, has since been applied more widely and controversially to explain various social interactions between humans, from the different mating strategies of males and females to the evolution of language.

'That one can use the same set of tools to analyse a game of tennis, the decision of when to run for office, predator–prey relationships, how much to trust a stranger, and how much to contribute to a public good makes game theory one of the most important analytic tools available to all of the social sciences,' noted Ostrom in 2012.[83] Ostrom's list is missing one of the earliest applications of game theory, one with which it remains practically synonymous even today. While economists were still scratching their heads over von Neumann and Morgenstern's tome, the US military quickly saw its worth for honing nuclear strategy. Much of that honing would be done at the RAND Corporation, a global policy think tank based a block from the beach in Santa Monica, California, that would quickly become a veritable *Who's Who* of game theorists charged with 'thinking about the unthinkable'.

7

The Think Tank by the Sea

Gaming nuclear war

'What are we to make of a civilization which has always regarded ethics as an essential part of human life, and . . . which has not been able to talk about the prospect of killing almost everybody, except in prudential and game-theoretic terms?'

Robert Oppenheimer, 1960

The Soviet newspaper *Pravda* once branded the organization based at the pink and white stucco building the 'American academy for death and destruction'. Despite moving to more modern headquarters in 2003, RAND remains synonymous with the Cold War and the icy logic of nuclear deterrence.[1] At the peak of RAND's notoriety in the 1960s, a folk song recorded by Pete Seeger summed up the organization's reputation for cold-blooded strategizing:

> Oh, the Rand Corporation's the boon of the world,
> They think all day long for a fee.
> They sit and play games about going up in flames;
> For counters they use you and me, honey bee,
> For counters they use you and me . . . [2]

If any one man can be regarded as the founding father of the RAND Corporation, then that man would be Henry 'Hap' Arnold, the commanding general of the US Air Force during the Second World War. Arnold was an early believer in the importance of a powerful and independent air force and never stinted from hitting his enemy with everything at his disposal during the war. 'We must not get soft,' Arnold warned Stimson when he heard the secretary of war had doubts about the firebombing of Dresden. 'War must be destructive and to a certain extent inhuman and ruthless.' Arnold's only regret was that the area bombing of German cities was proceeding too slowly. He was impatient for his scientists to invent 'explosives more terrible and more horrible than anyone has any idea of'.[3] They would duly oblige. Arnold's nickname was attributed to his sunny demeanour: 'Hap' was short for 'Happy'.

Months before the end of the war, Arnold began to worry that the scientific expertise assembled to aid the American military would quickly disperse after the conflict was over. Something of a visionary, Arnold foresaw the advent of intercontinental ballistic missiles (ICBMs). 'Someday, not too distant,' he wrote in 1943, 'there can come streaking out of somewhere – we won't be able to hear it, it will come so fast – some kind of gadget with an explosive so powerful that one projectile will be able to wipe out completely this city of Washington.'[4]

Arnold urged the Air Force to prepare for a future in which scientists would play a leading role in warfare. 'For the last twenty years we have built and run the Air Force on pilots,' he told the Navy top brass. 'But we can't do that anymore. We've got to think of what we'll need in terms of twenty years from now.'[5] On 7 November 1944, he wrote to his chief scientific adviser: 'I believe the security of the United States of America will continue to rest in part in developments instituted by our educational and professional scientists. I am anxious that the Air Force's post war and next war research and development be placed on a sound and continuing basis.'[6]

The man to whom his memo was addressed was none other than Theodore von Kármán, the same aerospace engineer that Max von Neumann had asked, unsuccessfully, to prevent his son devoting his life to the fruitless pursuit of mathematics. In the midst of the bloodiest war in history, Arnold urged his adviser, now a US citizen and director of the California Institute of Technology's aeronautical institute, to devote himself entirely to the task of investigating 'all the possibilities and desirabilities for post war and future war's development'.[7] 'I told these scientists', Arnold wrote in 1949, 'that I wanted them to think ... about supersonic-speed airplanes, airplanes that would move and operate without crew, improvements in bombs ... defenses against modern and future aircraft ... communication systems ... television ... weather, medical research, atomic energy.' In short, anything that 'might affect the development of the airpower to come'.[8]

Thirteen months later, von Kármán and his colleagues presented Arnold with a massive thirty-three-volume report entitled *Toward New Horizons*.[9] It did not disappoint him. 'The scientific discoveries in aerodynamics, electronics and nuclear physics open new horizons for the use of air power,' von Kármán wrote. His introduction preceded hundreds of pages of remarkably foresighted technical analysis, charting the way to developments such as intercontinental ballistic missiles and drones. Much of the information was culled from captured German scientists. The seeds of what would become the RAND Corporation appeared in a small section on the application of science to operations analysis – the brains of the war machine. The United States had assiduously developed expertise in mission-planning during the war. Ending that work, the report warned, would be 'a great mistake'. Instead, there should be established 'in peacetime a nucleus for scientific groups such as those which successfully assisted in the command and staff work in the field during the war. In these studies experts in statistical, technical, economic and political science must cooperate.'

Toward New Horizons was everything Arnold was expecting and more. But he had a war to win. He would not act on the report's recommendations until Frank Collbohm arrived in his office in September 1945. Collbohm was a tough, fit former test pilot and engineer who had joined the Douglas Aircraft Company in 1928. At the end of the war, Douglas was America's largest aircraft-maker, and Collbohm was the right-hand man of Donald Douglas, the company's founder. Collbohm had Arnold's ear too. During the war, he had alerted the general to the cutting-edge radar systems being developed at MIT. Later, he was asked to improve the performance of the B-29 bomber in raids over Japan. The team he helped lead calculated that stripping away most of the aeroplane's armour plating and leaving only a tail-gun would allow the B-29 to fly further and faster with a bigger payload. The Air Force accepted the recommendations, and the modified B-29s firebombed Japan's cities at a fiercer pace. Collbohm understood as well as anyone that science would play a pivotal role in any future conflict and was dismayed to see scientists drifting back to their universities at the end of the war.

When, a few weeks after the surrender of Japan, Collbohm met Arnold in Washington, D.C. to air his concerns, the general cut him off. 'Frank, I know what you're going to say,' he exclaimed, slamming his hand down on the table. 'It's the most important thing we have to do right now.'[10] Collbohm had also come with a proposal from Douglas himself: his company would be willing to house an independent group of scientists who would assist the Air Force with weapons research. Arnold liked the idea. He was good friends with Don Douglas, and a couple of years earlier, his son had married Douglas's daughter. Unperturbed by possible conflicts of interest, Arnold told Collbohm to get Douglas and meet him for lunch at Hamilton Field, an Air Force base just north of San Francisco on the opposite coast, in two days' time.

Collbohm caught the first plane he could – a B-25 bomber – to company headquarters in Santa Monica, where he rounded up Douglas and a few other company executives. The deal was done quickly. Arnold told the group he had $10 million of unspent funds from his wartime research budget which he was willing to give Douglas Aircraft to fund the new outfit. Douglas agreed to find space for the organization at their Santa Monica offices. Arthur Raymond, Douglas Aircraft's chief engineer, suggested the name: RAND for 'Research ANd Development'. Collbohm volunteered to lead it until a more suitable candidate could be found. His 'temporary' appointment as director would last twenty years. No tangible weapon would ever emerge from the think tank, only a slew of reports, prompting the joke that 'Research and *No* Development' was a more appropriate moniker. Some Army chiefs would forever think of the place as a coterie of 'pipe-smoking, trees-full-of-owls' intellectual types.

Project RAND was officially born on 1 March 1946, at the stroke of a pen on an Air Force contract. The terms specified the cash was for 'a continuing program of scientific study and research on the broad subject of air warfare with the object of recommending to the Air

Force preferred methods, techniques and instrumentalities for this purpose'. Initially, the scientists and mathematicians hired by RAND worked on technical projects ranging from nuclear propulsion to the design of new aircraft. The think tank's very first report was released on 2 May 1946. *Preliminary Design of an Experimental World-Circling Spaceship* concluded that 'modern technology has advanced to a point where it now appears feasible to undertake the design of a satellite vehicle'. Such a craft would be 'one of the most potent scientific tools of the Twentieth Century' and the achievement 'would inflame the imagination of mankind, and would probably produce repercussions in the world comparable to the explosion of the atomic bomb'. Eleven years later the Soviet Union put Sputnik into orbit, humbling the United States – and turbocharging both the space race and the arms race.

RAND's relationship with Douglas Aircraft quickly soured. Douglas complained that the Air Force was unfairly awarding contracts to its competitors to avoid accusations of favouritism. Meanwhile, RAND's growing cadre of mathematicians and social scientists felt straitjacketed by the firm's stiff corporate ethos. 'Academic people', commented Collbohm's fifth hire, the astronomer John Williams, 'have irregular habits and have never taken kindly to the eight-to-five routine.'[11] There was even resistance to the idea of ordering blackboards and chalk (which the academics wanted in four different colours). The break with Douglas Aircraft came on 14 May 1948, when, to the satisfaction of both parties, 'Project RAND' became the RAND Corporation, an independent, non-profit organization employing more than 200 people.

RAND's focus on engineering and physics would soon be widened enormously under Williams's stewardship. He was recruited in 1946 on the advice of Collbohm's close friend Warren Weaver, the wartime director of the Applied Mathematics Panel (AMP).[12] The AMP had carried out during the war exactly the kind of research that Collbohm thought RAND should be doing in peacetime.

Weaver was himself a former mathematics professor but had little time for the 'dreamy moonchildren, the prima donnas, the asocial geniuses' that he felt dominated the higher echelons of academic science.[13] Williams was another of Weaver's practically minded recruits.

After graduating from the University of Arizona in 1937, he started a PhD in astronomy at Princeton University but became so busy with war work that he never finished.

During the Second World War, the AMP supported the new field of 'operations research', pioneered in Britain by the physicist Patrick Blackett. Operations research brought the methods of the sciences to bear on wartime problems. The idea was simple: collect and analyse as much data as possible, test hypotheses in the field and use the results to home in on solutions. In nine short months after he joined Coastal Command as a science adviser in 1941, Blackett used this approach to turn around the Royal Air Force's unsuccessful campaign against German U-boats. Blackett and his team calculated, for example, that setting the Air Force's depth charges to explode at a depth of 25 feet rather than 100 feet (as they had been) would result in two and a half times more hits. The change was so effective that captured U-boat crews thought the British had started using a new and more powerful explosive.[14] The idea caught on quickly. By the end of the war, around 700 scientists in the United States, Canada and Britain were employed in operations research.

With the war over, the question facing military planners was: how could all that expertise now be put to use? With defence budgets tightening, spending on new weapons systems or military operations would have to be weighed carefully against other demands. Weaver's solution to this problem was the notion of 'military worth', a simple score that captured all the complex pros and cons of such choices so that decisions could be made more easily. And the mathematical apparatus to carry out military worth calculations was that of game theory. 'Military worth, as the phrase is here used, is closely related to the general concept of *utility* in economic theory,' Weaver explained in a report in 1946, referring his reader to the relevant section of von Neumann and Morgenstern's *Theory of Games*. 'This pioneering and brilliant book is, it should be pointed out, connected in a most important way with the viewpoint here being presented, for it develops a large part of the mathematics necessary for theories of competitive processes.'[15]

In September 1947, at a RAND-sponsored conference in New York, Weaver set out a manifesto for the nascent organization.

Operations research had 'resulted only from the pressure and necessity of war', he said. RAND would provide in peacetime an environment where similar techniques could be more widely used for 'analyzing general theories of warfare'. Chess master Emanuel Lasker's Jazz Age dreams of a 'science of contest' were at last taking shape. 'I assume that every person in this room is fundamentally interested in and devoted to what can broadly be called the rational life . . . as compared with living in a state of ignorance, superstition and drifting-into-whatever-may-come,' Weaver continued.

> I think that we are not interested in war but in peace . . . I assume that every person in this room is desperately dedicated to the ideals of democracy, and to so running our own business, so cleaning our own house, and so improving our own relations with the rest of the world that the value of those ideals in which we believe becomes thereby evident.[16]

RAND analysts pride themselves on their dedication to the 'rational life' to the present day. The organization's commitment to peace and democracy – at least beyond the borders of the United States – would be brought into question again and again.

When Weaver's protégé Williams was hired, the new section at RAND he headed was devoted to the 'Evaluation of Military Worth'. An aficionado of game theory himself, Williams would write a humorous primer on the subject, *The Compleat Strategyst*, strewn with in-jokes and featuring many of RAND's analysts, transformed into comic characters. Translated into at least five languages including Russian, the book would become one of RAND's most popular publications.

Williams swiftly began recruiting experts in the field. In 1950, RAND's annual report would proclaim:

> the analysis of systems for strategic bombardment, air defense, air supply, or psychological warfare, pertinent information developed or adapted through survey, study or research by RAND is integrated into models, largely by means of mathematical methods and techniques . . . In this general area of research . . . the guiding philosophy is supplied by the von Neumann-Morgenstern mathematical theory of games.[17]

Williams would shape both the intellectual and physical environment of RAND for the next twenty years until his death, aged fifty-five, in 1964. He campaigned successfully for two new divisions to be created at RAND – one for social science, the other for economics. In 1953, the rapidly growing outfit moved to purpose-built quarters by the beach. The new building's lattices of courtyards and corridors, designed to increase chance meetings between staff from different divisions, were built to specifications drafted by Williams. In this, as in so many things, Williams would prove to be ahead of his time.

Rotund, weighing close to 300 pounds, Williams enjoyed the good life. He had RAND's machinists fit a Cadillac supercharger to the engine of his brown Jaguar, taking it out for midnight runs down the Pacific Coast Highway at more than 150 miles per hour. At his house in Pacific Palisades, the booze flowed so freely that his erudite guests would be rolling around on the floor drunk by the end of his parties. If that all sounds rather familiar, it is no coincidence. Williams had attended von Neumann's lectures at Princeton and worshipped him. Von Neumann's spirit suffused RAND from its inception. All that was missing was the great man himself. On 16 December 1947, Williams wrote to his old professor.

'The members of the Project with problems in your line (i.e. the wide world) could discuss them with you, by mail and in person,' von Neumann read. 'We would send you all working papers and reports of RAND which we think would interest you, expecting you to react (with frown, hint, or suggestion) when you had a reaction.' For his services, von Neumann would receive US$200 a month – the average monthly salary at that time. The offer from Williams came with a charming stipulation: 'the only part of your thinking time we'd like to bid for systematically is that which you spend shaving: we'd like you to pass on to us any ideas that come to you while so engaged'.[18]

Von Neumann began consulting for RAND the following year, holding court there much as he did at Los Alamos and Princeton. As he ambled through the criss-crossing corridors, people would call him aside to pick his brains about this or that. Williams would throw demanding maths problems at his hero in an effort to trip him up. He never succeeded. At one of his 'high-proof, high-I.Q. parties'[19] one analyst produced a fat cylindrical 'coin' that was something of a RAND obsession at the time. Milled by the RAND machine shop at the behest of Williams, their proportions were carefully chosen so that the chances of falling heads, tails or on its side were equal. Without blinking an eye, von Neumann correctly stated the coin's dimensions.[20]

Like RAND's analysts, von Neumann was fascinated by war strategy. As a child he'd played the eighteenth-century game Kriegspiel with his brothers, drawing terrain for battles on graph paper, and he found that a version of the game was popular during lunchbreaks at RAND. He was also familiar with the idea of 'military worth' and had helped forge its links to game theory. On 1 October 1947, just a couple of months before he received the letter from Williams, the statistician George Dantzig had paid him a visit. Dantzig, a former liaison between the Air Force and AMP, wanted to solve the daunting logistical problem of matching the military's needs to the resources available as quickly and efficiently as possible. Air Force budgeting in the 1940s was so baroque that producing a plan to requisition the appropriate manpower and materiel for a task could take seven months or more. Eventually Dantzig would help to invent an entirely new discipline called 'linear programming' to deal with the process. But in 1947, he had begun with a relatively simple objective: to devise a diet that met a soldier's nutritional needs as cheaply as possible.[21] The numbers involved in even this supposedly straightforward problem had, however, spiralled rapidly out of control, and he had decided to ask von Neumann, expert in computing techniques, for help. Dantzig, who would join RAND in 1952, had begun describing the matter in detail when, uncharacteristically rudely, von Neumann impatiently told him to 'Get to the point.' Annoyed himself, Dantzig then 'slapped the geometric and the algebraic version of the problem on the blackboard' in 'under one minute'. Dantzig recalls what happened next:

> Von Neumann stood up and said, 'Oh that!' Then for the next hour and a half, he proceeded to give me a lecture on the mathematical theory of linear programs.
>
> At one point, seeing me sitting there with my eyes popping and my mouth open – after all I had searched the literature and found nothing,

> von Neumann said: 'I don't want you to think I am pulling all this out of my sleeve on the spur of the moment like a magician. I have just recently completed a book with Oscar Morgenstern on the theory of games. What I am doing is conjecturing that the two problems are equivalent. The theory that I am outlining for your problem is the analog of the one we have developed for games.'[22]

Von Neumann had instantly recognized that Dantzig's optimization problem was mathematically related to his minimax theorem for two-person zero-sum games. The insight helped determine the conditions under which logistical problems of the type Dantzig was interested in could or could not be solved. Linear programming is now a staple approach to such problems – from the placement of servers inside data centres to the purchase and distribution of vaccines.

The twin influences of these military mathematicians and the Air Force meant that RAND's interests in 1948 were completely aligned with von Neumann's three main obsessions of the time: computing, game theory and the bomb. For a while, there was no other setting that von Neumann enjoyed more and for the next few years, until his interests diverged, von Neumann would often visit the Santa Monica think tank. Even when he was not physically present, his influence was felt. 'Everybody knew that von Neumann was king,' recalled Jack Hirshleifer, who was employed in the economics division.[23]

Early computing work on the 'Super' required random numbers for Monte Carlo simulations, so RAND engineers built an electronic device to generate them. This was compiled into a surprise best-seller entitled *A Million Random Digits and 100,000 Normal Deviates*. In 1949, Williams headed a RAND team that visited various firms to gauge their ambitions for developing electronic computers. 'It was a dismal scene,' Williams complained in a memo after he found out that their plans were non-existent.[24]

RAND turned to von Neumann, by now considered the foremost expert on computing in the United States. With his tongue firmly in his cheek, von Neumann questioned whether a computer was needed at all. According to journalist Clay Blair, RAND scientists came to him with a problem they thought too difficult to solve by conventional means:

> After listening to the scientists expound, Von Neumann broke in: 'Well, gentlemen, suppose you tell me exactly what the *problem* is?'
>
> For the next two hours the men at Rand lectured, scribbled on blackboards, and brought charts and tables back and forth. Von Neumann sat with his head buried in his hands. When the presentation was completed, he scribbled on a pad, stared so blankly that a Rand scientist later said he looked as if 'his mind had slipped his face out of gear,' then said, 'Gentlemen, you do not need the computer. I have the answer.'
>
> While the scientists sat in stunned silence, Von Neumann reeled off the various steps which would provide the solution to the problem. Having risen to this routine challenge, Von Neumann followed up with a routine suggestion: 'Let's go to lunch.'[25]

RAND pressed ahead with building their own machine – piggybacking on von Neumann's computer project at the IAS. A team from RAND travelled to Princeton to learn from the IAS experience and, like others around the world, they eagerly read Goldstine and von Neumann's updates. In 1952, RAND hired Willis Ware, an electrical engineer who had worked on the IAS machine from 1946 to 1951. He stayed for fifty-five years, serving as head of RAND's Computer Science department from 1960.

RAND's machine started work in 1953, often running Monte Carlo bomb simulations and Dantzig's logistical problems. They called it the JOHNNIAC (John von Neumann Numerical Integrator and Automatic Computer). A framed photograph of the man himself hung on the wall next to the machine.

To begin with, von Neumann's energies were focused on deepening the mathematics of game theory at RAND. A letter from Williams in December 1947 promised that his department planned to make 'major efforts on applications of game theory'. Von Neumann's response is encouraging. 'The work on game theory, which you have been pushing so energetically and successfully interests me greatly,' he wrote back. 'I don't think that I need tell you this again.'[26] Von Neumann reviewed the work of RAND's mathematicians on the subject, and his own early publications for the organization looked at solutions of two-person and *n*-person games. Like other game theorists there, von Neumann was now less interested in proving new theorems about

The JOHNNIAC.

game theory than coming up with ways to compute actual solutions. 'I have spent a good deal of time lately on trying to find numerical methods for determining "optimum strategies" for two-person games,' von Neumann reported to Weaver in March 1948. 'I would like to get such methods which are usable on an electronic machine of the variety which we are planning, and I think that the procedures that I can contemplate will work for games up to a few hundred strategies.'

One particularly fertile and long-lived area of research at RAND would be the mathematics of 'duels' – the subject of nearly a hundred papers and memoranda over two decades. In RAND, the duel served as a simplified model for diverse situations: two aircraft or two tanks closing in for combat, for instance, or a bomber versus a battleship. For analysts, the duel allowed the relatively complete maths of the two-person zero-sum game to be brought to bear on real combat data from the Second World War. In the duels RAND considered, each player wanted to get the best shot by holding off firing as long as possible – but still sooner than their opponent. RAND mathematicians explored many permutations of the duel. If the duellists heard each other's shots, the duel was 'noisy', if neither learned whether or when their opponent had pulled the trigger, the duel was 'silent'. Opponents might each have one bullet or many, one duellist could be a poorer shot than the other and so on. Each scenario was brought to a precise solution by dozens of RAND's visiting or resident scholars.[27]

Some of the papers, including 'Silent Duel' and 'One Bullet Versus Two, Equal Accuracy', were written by a mathematician called Lloyd Shapley, son of astronomer Harlow Shapley, one of the most famous scientists in America. The younger Shapley's mathematics degree at Harvard University had been interrupted by the war. He spent the next two years in China, helping the US Air Force decipher coded Soviet weather reports for the region – work that contributed to the forecasts for Japan. After returning to Harvard to finish his studies, and with no interest in postgraduate work, he joined the 'Military Worth' section under Williams after hearing about RAND through Air Force connections. He first came to von Neumann's attention in dramatic fashion during a packed RAND seminar in the summer of 1948.[28]

Von Neumann had been asked to prove that a particular duel – one involving two fighter aircraft – had no formal solution. After his customary minute or so of staring into space, von Neumann had raced away on the blackboard when he was suddenly interrupted by a voice from the back of the room.

'No! No! That can be done much more simply!'

Shocked silence settled over the room. Hans Speier had just been appointed head of RAND's new social science division. His memories of the incident were still fresh in his mind decades later:

> 'Now my heart stood still, because I wasn't used to this sort of thing,' said Speier.
>
> Johnny von Neumann said, 'Come up here, young man. Show me.' He goes up, takes the piece of chalk, and writes down another derivation, and Johnny von Neumann interrupts and says, 'Not so fast, young man. I can't follow.'
>
> Now . . . he was right, the young man was right . . . [29]

Von Neumann was astonished. 'Who is this boy?' he asked Williams after the meeting. It was Shapley, who Williams had hired earlier that year. 'And what', said von Neumann, 'has he been doing?'

'Only John Williams could do this marvellously,' Speier continued. 'He said, 'Oh well, he has written three or four papers, each of which is the equivalent of a doctoral dissertation in mathematics'.

Which was true. Johnny von Neumann looked at that, and he gave him – I don't know – it was something quite fantastic, a special stipend to Princeton or something like that.

With von Neumann's encouragement Shapley did indeed go to Princeton in 1950, by now a hotbed of game theory research. While he was there, he solved a key question raised – but not answered – by von Neumann and Morgenstern's cooperative game theory: how can payouts be divided 'fairly' among the members of a coalition? He defined a way to distribute the payouts, now called the Shapley values of a cooperative game, such that no player could do better either by themselves or in any possible splinter groups.

One way to calculate the Shapley values for a game is to imagine that costs or benefits are allocated by a committee on a first-come, first-served basis. Since the order in which players join a grand coalition should not matter in a truly 'fair' settlement, the Shapley value

Lloyd Shapley at RAND in the 1970s.

for a game actually comprises the payouts that each player gets after averaging over all possible orders in which the players might approach the committee.[30]

There is something magical in the way the Shapley value elegantly solves what was intractable to the authors of *Theory of Games*. This was the first hint that the cooperative game theory of von Neumann and Morgenstern could solve real-world problems. And Shapley would do much more. One of his most influential contributions would be forged with fellow Princeton mathematician David Gale, who loved mathematical games and conundrums. Gale was known for quietly scribbling down a grid or pulling a fistful of coins from his pocket at mealtimes and challenging fellow graduate students to solve some puzzle or other. Perhaps in this spirit, Gale sent a few other mathematicians a note in 1960 asking, 'For any pattern of preferences, is it possible to find a stable set of marriages?'

Shapley sent his answer by return of post.

> Let each boy propose to his best girl. Let each girl with several proposals reject all but her favorite, but defer acceptance until she is sure no one better will come her way. The rejected boys then propose to their next-best choices, and so on, until there are no girls with more than one suitor. Marry. The result is stable, since the extramarital liaisons that were previously rejected will be disliked by the girl partners, while all others will be disliked by the boy partners.[31]

Shapley's solution applies to *any* market that requires two sets of people to be paired up so that no one can do better. Gale and Shapley wrote up the findings in a paper, in which they also showed that the method could be used to match applicants to colleges.[32] Now known as the Gale-Shapley 'deferred acceptance' algorithm, the work was cited as part of the decision to award Shapley a share of the 2012 Nobel Memorial Prize in Economic Sciences, instantly garnering him a reputation in the press as a mathematical matchmaker extraordinaire. He would be one of two game theorists whose striking early contributions did much to make the field useful for economics and other disciplines. At Princeton, the twenty-six-year-old Shapley met the other – a graduate student five years his junior called John Nash.

Shapley was cultured, popular, a virtuoso piano player and a decorated war hero. He was also an expert player of Kriegspiel and Go. Nash promptly fell in love with him.

Most people who have heard of John Nash know of him through Ron Howard's *A Beautiful Mind*, which is unfortunate because the film romanticizes the man and somewhat mischaracterizes his work. The Nash portrayed in Sylvia Nasar's biography, upon which the film is nominally based, is a petulant bully who advises his mistress of four years to give up their son for adoption. The physically imposing Nash later threw his future wife 'to the ground and placed his foot on her neck' during a maths department picnic.[33]

When he met Shapley, 'Nash acted like a thirteen-year-old having his first crush,' writes Nasar.

> He pestered Shapley mercilessly. He made a point of disrupting his beloved Kriegspiel games, sometimes by sweeping the pieces to the ground. He rifled through his mail. He read the papers on his desk. He left notes for Shapley: 'Nash was here!' He played all kinds of pranks on him.[34]

It was to be one of several emotionally tempestuous relationships that Nash had with other men.

Postwar Princeton was a town awash with genius, but Nash implied that he was smarter and from superior stock than his fellow students – particularly those from Jewish backgrounds. 'He definitely had a set of beliefs about the aristocracy,' says fellow Princetonian Martin Davis. 'He was opposed to racial mixing. He said that miscegenation would result in the deterioration of the racial line. Nash implied that his own blood lines were pretty good.'

Still Shapley humoured him, dazzled by the younger mathematician's obvious brilliance. 'Nash was spiteful, a child with a social IQ of 12, but Lloyd did appreciate talent,' recalled Shapley's roommate, economist Martin Shubik, whom Nash never called anything other than 'Shoobie-Woobie'. Together with fellow graduate student John McCarthy, Shubik, Shapley and Nash invented a board game in which players had to form alliances – then break them at the last possible minute to win. The game, later named 'So Long, Sucker', was designed to push people to their psychological limit: some married couples reputedly went home in separate cabs after a night of play. Once, after Nash ditched McCarthy particularly ruthlessly to win, McCarthy exploded. 'But I didn't need you anymore,' a bemused Nash kept saying. Nash's name for the game? 'Fuck your buddy!'

Like Shapley, Nash was drawn to game theory. He attended the popular weekly seminars held on the subject by Albert Tucker, chair of the Princeton maths department. One of the first speakers was von Neumann. It was in these seminars that Nash developed the ideas behind his first game theory paper, written up with the encouragement of Morgenstern, whom Nash referred to as Oskar La Morgue behind his back. Impressively, Nash had first thought of his approach to the problem as an undergraduate at the Carnegie Institute of Technology, during the only economics class he would ever attend. In 'The Bargaining Problem', Nash showed that a two-person cooperative game could be brought to a solution if certain conditions were met.[35] He attacked the problem with the axiomatic method championed by Hilbert – and, of course, von Neumann. According to orthodox economic theory, there is no unique solution to how two parties should divide up the 'surplus' created when they strike a deal. Von Neumann and Morgenstern got no further with the problem either: they could only give a spectrum of solutions for the two-person game, not a single point at which the players would come to an agreement. Only the simplest 'symmetric' version of the problem – when the parties have exactly the same interests and bargaining power – had been shown to have an exact solution. In this case, the net gain would obviously be split equally between the two. Nash was able to show more generally that even in asymmetric cases, assuming that the utilities could be assigned to the two parties in the way *Theory of Games* described, the bargaining problem has an exact solution – namely, the point at which the product of the two utility scores is a maximum. The result prefigured in a small way Shapley's contribution, which provides a similar sort of solution for cooperative games with any number of players.

The young Nash had never lacked confidence. In 1948, his very first year as a graduate student, he arranged to see Einstein in his office at the IAS to discuss some pressing ideas on the interaction of particles with fluctuating gravitational fields. Nash spent nearly an

John Nash. *Courtesy MIT Museum.*

hour trying to unwind his thoughts at Einstein's blackboard but eventually came unstuck. 'You had better study some more physics, young man,' Einstein told Nash with a kind smile before sending him on his way. So the following autumn, when the unabashed Nash thought he had made a breakthrough in game theory, he quite naturally scheduled a meeting with the discipline's founding father.

Just as he had with Einstein, Nash arranged to see von Neumann in his office at the IAS. In 1949, von Neumann was busy consulting for the government, the military, big business and RAND. Behind the scenes, he was campaigning for America to pursue the H-bomb while madly chasing the computing resources necessary to show it was possible. He was also building his own computer at the IAS. An apologetic letter to a friend dictated to his secretary, Louise, that spring read, 'I am delayed by a siege of work, which I hope will last only for a few days or so. At this point there was a considerable burst of hilarity from Louise. Can you interpret it?' Still, he made time to see the promising young graduate student.

Nash nervously started to present what would turn out to be his greatest – and final – contribution to game theory. He had come up with a mathematical framework allowing the analysis of any type of game – whether zero-sum or not – with any number of participants, and showed that there are certain outcomes for all games in which no player can do any better by unilaterally changing their strategy. These kinds of solutions to a game are now called Nash equilibria. It was a staggering accomplishment, though no one, least of all Nash, had any idea how thoroughly useful his idea would prove to be. There was one catch: in Nash's scheme, players were not allowed to communicate or team up – almost as if they were caught in a perpetual final round of 'Fuck your buddy!' Von Neumann hated it.[36] When Nash started describing his proof, von Neumann interrupted, finished off his chain of reasoning and shrugged away Nash's accomplishment with a devastating denouement: 'That's trivial, you know,' he said. 'That's just a fixed-point theorem.' And it was. Nash had used the same elegant trick that von Neumann had in his 1937 model of an expanding economy. Von Neumann was underwhelmed, as other mathematicians would be. They would regard Nash's proof as solid enough, and a great PhD thesis, but not a patch on his later work on nonlinear partial differential equations, for instance, which would net him the Abel Prize in 2015.

What von Neumann disliked most about Nash's approach, though, was the axioms upon which it was built. The idea that people might not work together for mutual benefit was anathema to him. He was central European to the core, his intellectual outlook shaped by a milieu where ideas were debated and shaped over coffee and wine. At that very moment, he was busily pushing as much as he could about the technical details of his computing project into the public domain. Ruling out communication ran counter to the spirit of his 'coalitional' conception of game theory. Even the fact that Nash's solution could produce a point-solution to a complex game struck von Neumann as unrealistic. He maintained the theory could provide only a spectrum of solutions, with the actual result determined by social mores and the specific circumstances at the time.

Nash would later ascribe von Neumann's coolness as a defensive response prompted by a Young Turk invading his turf. 'I was playing a non-cooperative game in relation to von Neumann rather than simply seeking to join his coalition,' he told the historian Robert Leonard. 'And of course, it was psychologically natural for him not to be entirely pleased by a rival theoretical approach.'[37] 'Natural', perhaps, from Nash's point of view and in keeping with reports that von Neumann could react angrily to being contradicted.[38] But von Neumann's rather more magnanimous reaction to being corrected brusquely, and publicly, by Shapley the year before at RAND suggests there was more at stake for him here than the embarrassment of being outfoxed by a younger mathematician.

A few days after Nash's meeting with von Neumann, he found a more sympathetic ear. 'I think I've found a way to generalize von Neumann's min-max theorem,' he told Gale. 'And it works with any number of people and doesn't have to be a zero-sum game.' Gale urged him to rush the result into print, and he helped draft a paper based on Nash's proof.[39] 'I certainly knew right away that it was a thesis,' Gale told Nasar. 'I didn't know it was a Nobel.'[40]

Others were quicker to spot the potential of Nash's work. By the time he had finished his thesis the following year, he had an offer of a permanent job at RAND. Nash declined, preferring to find a faculty position with the greater intellectual freedom that entailed but elected to spend his summers at the Santa Monica think tank as a consultant instead. His relationship with the think tank would only come to an end in 1954, when he became a victim of one of the many police sting operations that aimed to force gay men to leave town. Nash was caught exposing himself in a public lavatory in the early hours of the morning. When RAND's head of security came to see him the next morning, he denied he was homosexual and instead told him he had been conducting an 'experiment'. 'I like women,' he insisted, producing a photo of his mistress and illegitimate son. Nash was promptly escorted from the building, and his security clearance revoked.[41]

Nasar contends that von Neumann's rejection stung Nash so badly that 'he never approached von Neumann again'. Yet whatever rift existed between the two men did not prevent them from attending game theory workshops together at RAND. Nor did it stop von Neumann from directing his readers to Nash's work on non-cooperative games in the 1953 preface to the third edition of *Theory of Games*. By

1955, relations were cordial enough for von Neumann to chair a presentation by Nash on his 'Opinions on the Future Development of Theory of *n-person* Games' on the last day of a conference held at Princeton University.[42] Nash argued again that the theory produced too many solutions for most games. Von Neumann, again, politely demurred.

Surveys of the roots and influence of game theory have generally taken a dim view of its progenitor. 'Game theory portrays a world of people relentlessly and ruthlessly but with intelligence and calculation pursuing what each perceives to be his own interest,' says the physicist turned historian Steve J. Heims. 'The harshness of this Hobbesian picture of human behaviour is repugnant to many, but von Neumann would much rather err on the side of mistrust and suspicion than be caught in wishful thinking about the nature of people and society.'[43]

Heims attributes von Neumann's misanthropy to his experience of living under Béla Kun's regime as a teenager in Hungary. But what von Neumann saw happen in Germany scarred him very much more. 'His hatred, his loathing for the Nazis was essentially boundless,' says Klári. 'They came and destroyed the world of this perfect intellectual setting. In quick order they dispersed the concentration of minds and substituted concentration camps where many of those who were not quick enough . . . perished in the most horrible ways.'[44]

By the time von Neumann visited Europe again in 1949, his belief in people had evaporated away altogether. 'I feel the opposite of nostalgia for Europe,' he wrote to Klári, 'because every corner reminds me . . . of the world which is gone, and the ruins of which is no solace. My second reason for disliking Europe is the memory of my total disillusionment in human decency between 1933 and September 1938.'[45]

Still, in *Theory of Games*, arch-rationalist von Neumann had presupposed that even the hard-boiled players he envisaged would collaborate for common advantage. By contrast, Nash would himself describe his thinking, in retrospect, as more individualistic, more 'American'.[46] It is arguably Nash's conception of game theory, not von Neumann's, that more closely embodies the kill-or-be-killed paranoia of the early Cold War. And it would be Nash's powerful solution to games that would for the first few decades after the Second World War take academia, economics and RAND by storm.

In Santa Monica, RAND's analysts had found themselves bumping up against the limits of game theory. When the mathematics could not provide an answer to a problem, they turned to experimenting – frequently on each other. In the summer of 1949, before Nash's work had been published, mathematician Merrill Flood had started exploring how well the theory predicted human behaviour by devising games and dilemmas to test his fellow RANDites. Sometimes he turned an everyday problem into a bargaining 'game' to see if he could find a 'rational' solution. Flood published some of his research in a RAND memorandum entitled 'Some Experimental Games'.[47] The results of his investigations were often surprising. Things did not always go to plan.

In June that year, Flood had tried to buy a used Buick from his friend and colleague, the futurist and nuclear strategist Herman Kahn, who was planning to move back east with his family. Flood recast the situation as a sort of two-player game where the object was to reach an agreement on a fair price for the car. By avoiding a car dealer, they could split the 'surplus' profit between themselves. The two men decided to consult a used-car dealer they both knew and were able to ascertain his buying and selling price. Having established the dealer's 'cut', they were now left to agree on how they should divide the sum 'fairly' – the classic bargaining problem that Nash had addressed in his paper. Flood suggested an even split: he would pay Kahn the dealer's buying price plus half of what the dealer would have made by selling the car himself. But both of them knew any other division of the profits was equally admissible – and rational. In the end Kahn, probably fed up with the circularity of the discussion, drove his Buick to the east coast without completing the sale.

News of Nash's discovery spread to RAND later that year. With two equally valid solutions to choose from, Nash or von Neumann-Morgenstern, the question was which would real humans plump for? In January 1950, Flood and his colleague Melvin Dresher performed an experiment to find out. Transformed into an anecdote by Princeton's Tucker, who was also a RAND consultant, it would become the most notorious 'game' to emerge from the theory: the Prisoner's Dilemma.

Tucker's spin on the experiment was to present Flood and

Dresher's game as a choice facing two prisoners being held separately by police. The tale has been refined over the years and now usually runs something like this:[48]

> Two members of a criminal gang are arrested and imprisoned. Each prisoner is in solitary confinement with no means of communicating with the other. The prosecutors lack sufficient evidence to convict the pair on the principal charge, but they have enough to convict both on a lesser charge. Simultaneously, the prosecutors offer each prisoner a bargain. Each prisoner is given the opportunity either to betray the other by testifying that the other committed the crime, or to cooperate with the other by remaining silent. The possible outcomes are:
>
> If A and B each betray the other, each of them serves two years in prison.
>
> If A betrays B, but B remains silent, A will be set free and B will serve three years in prison.
>
> If A remains silent but B betrays A, A will serve three years in prison, and B will be set free.
>
> If A and B both remain silent, both of them will serve only one year in prison (on the lesser charge).

The only Nash equilibrium of the dilemma is for the prisoners to rat on each other. To see why, imagine you are prisoner A. If you betray B, and he does the same to you, you both get two years inside. On the other hand, if he stays quiet, you get off scot free. If you refuse to turn stool pigeon, you get a three-year sentence if he betrays you but even if he also opts to stay silent, you still serve a year in prison. Betrayal is the better option no matter what your partner decides – though if both prisoners talk to the police, the outcome is worse than if they both do not.

So much for the 'rational' choice. Flood and Dresher wondered what real players would choose to do in this non-zero-sum game.

> We conducted one brief experiment with a two-person positive-sum non-cooperative game in order to find whether or not the subjects tended to behave as they should if the Nash theory were applicable, or if their behavior tended more toward the von Neumann-Morgenstern solution, the split-the-difference principle, or some other yet-to-discovered principle.

Prisoner A \ Prisoner B	Stay Silent	Rat
Stay Silent	1 year, 1 year	3 years, 0 years
Rat	0 years, 3 years	2 years, 2 years

The Prisoner's Dilemma. Payoffs are listed in the order Prisoner A, Prisoner B.

Their experiment – dubbed 'A Non-cooperative Pair' – presented the players with two strategies each and the payoffs table below. Flood and Dresher roped in Williams (player 'W'), head of their section, and economist Armen Alchian (player 'A'). Both were reasonably familiar with zero-sum two-person games but knew nothing of Nash's proposed solution for non-zero sum games like the Prisoner's Dilemma. The game progressed for a hundred rounds, and the players were asked to jot down their reactions and reasoning during play.

The experiment is particularly vicious because the winnings are heavily skewed in favour of player W. If the players cooperate, player A gets a half cent, player W gets a cent. If they both choose to defect, player A gets nothing, but player W still wins a half cent. The Nash

Player A \ Player W	Strategy 1 Defect	Strategy 2 Cooperate
Strategy 1 Cooperate	-1 cent, 2 cents	½ cent, 1 cent
Strategy 2 Defect	0, ½ cent	1 cent, -1 cent

A non-cooperative pair. Payoffs listed in the order Player A, Player W.

equilibrium is in the bottom-left corner square – both players should defect. Had they played this strategy throughout the match, Williams would have ended the game with 50 cents and Alchian with nothing. In the end, Alchian came away with 40 cents and Williams won 65 cents. The two cooperated in sixty out of 100 plays – much more often than a 'rational' player should. 'It seems unlikely that the Nash equilibrium is in any realistic sense the correct solution,' Flood notes.[49] Though the participants were prohibited from reaching an understanding on dividing up the winnings, they leaned towards the von Neumann-Morgenstern solution of mutual cooperation.

The two players' notes reveal their thinking. Williams quickly worked out that defecting in every game would net him at least a half cent per round – but that with 'nominal assistance from AA', they would both be better off. 'This means I have control of the game to a large extent, so Player AA had better appreciate this and get on the bandwagon.'[50]

Alchian did not see things entirely this way. He starts off assuming that Williams will simply defect, as this is his path to a 'sure win'. When Williams cooperates, he is thrown into confusion: 'What is he doing?!!' After playing 'defect' a few more times, and getting 'defect' back on the next turn from Williams, he eventually responds to Williams' bid for cooperation by cooperating in turn. But later he appears to grow resentful that Williams gains more through mutual cooperation than he does. 'He will not share,' Alchian complains repeatedly. He tries to level things up and plays 'defect' a few times with the expectation that Williams will continue to cooperate. When Williams retaliates with 'defect', Alchian cooperates for much of the remaining game.

Flood and Dresher canvassed Nash's opinion on their experiment. He responded by poking holes in their methodology. 'The flaw in the experiment as a test of equilibrium point theory is that the experiment really amounts to having the players play one large multi-move game,' he wrote. 'One cannot just as well think of the thing as a sequence of independent games as one can in zero-sum cases. There is much too much interaction . . .' He could not resist taking a swipe at the players. 'It is really striking however how inefficient AA and JW were in obtaining the rewards,' he said, adding snidely: 'One would have thought them more rational.'

Nash was right that the experimental conditions are far from an ideal test of his theory. The problem is that the Nash equilibrium for the 100-move game is for players to defect every time. To see why, imagine that the players are about to play the last round of the game. A rational player should seize the chance to secure a bigger payoff by defecting, since her opponent will not be able to retaliate. But then knowing that her opponent will reason the same way and defect, it is also logical for her to defect in the penultimate round, and so on.[51] Yet even the hyper-rational, amoral denizens of RAND did not do this. Flood recalls that von Neumann was quite tickled by their experiment. As he had predicted, players did not naturally gravitate towards the Nash equilibrium.[52] Beyond that, however, he seems to have taken remarkably little interest.

The Prisoner's Dilemma is often portrayed as a paradox of rationality because the most rational course of action for each individual adds up to a worse outcome for everyone. Flood and Dresher hoped von Neumann would 'solve' the Prisoner's Dilemma. He did not. Neither has anyone else – though a lot of ink has been spilt in the effort to do so. Most game theorists now agree there is no 'answer' to the dilemma of the sort that Flood and Dresher envisaged. That is because there is no real paradox. The incentives of the game ensure that rational players do not cooperate. The real mystery is why humans faced with a one-shot Prisoner's Dilemma still sometimes do.[53]

By 1946, von Neumann was predicting that devastating nuclear war was imminent. 'I don't think this is less than two years and I do think it is less than ten,' he wrote to Klári on 4 October that year.[54] His answer was preventive war – a surprise attack that would wipe out the Soviet Union's nuclear arsenal (and a good number of its people too) before the country was able to retaliate. 'If you say why not bomb them tomorrow, I say why not today?' he reportedly said in 1950. 'If you say today at 5 o'clock, I say why not one o'clock?'[55]

Some have argued that von Neumann viewed the stand-off between the superpowers as a Prisoner's Dilemma[56] or that his (ultimately unfounded) fears of an imminent Third World War were rooted in game theory. 'There was perhaps an inclination to take a too exclusively rational point of view about the cases of historical events,' Ulam noted in his obituary of von Neumann. 'This tendency was possibly due to an over-formalized game theory approach.'[57]

There is no evidence that von Neumann *explicitly* viewed the Cold War or the arms race in these terms. He hated communist Russia and in the wake of the deadliest war that the world had ever known one did not have to be a game theorist to favour preventing another – even by the brutal means of an atom-bombing campaign that left millions of Russians dead. Wigner, who perhaps knew and understood von Neumann better than anyone, put it differently. 'A mind of von Neumann's inexorable logic,' he wrote, 'had to understand and accept much that most of us do not want to accept and do not even wish to understand.'[58]

For all its horror, pre-emptive nuclear war was a surprisingly popular idea in the higher echelons of power. Many in the US military were keen.[59] 'The only certain protection against aggression', Arnold told Secretary of War Henry Stimson in 1945, 'is to meet it and overcome it before it can be launched or take full effect.' But they were not alone. Supporters in the press included William Laurence, science correspondent of the *New York Times*, who wanted the US to force the Soviets to accept nuclear disarmament. If that led to war, Laurence argued, 'it would be to our advantage to have it while we are still the sole possessors of the atomic bomb'.[60] Laurence had more reason than most to fear a nuclear attack on America. He was the only journalist to have been invited to witness the Trinity test and had seen the destructiveness of the atom bomb at first hand.

Senior figures in the Truman and Eisenhower administrations pressed for nuclear strikes privately – and sometimes in public. Truman never seriously considered making an unprovoked attack on the Soviet Union, but Eisenhower did contemplate using atom bombs against China on several occasions during the 1950s. Since the Sino-Soviet Treaty of 1950 included provisions for mutual military assistance should one partner be attacked, Eisenhower knew that an attack on the Chinese would probably need to be coordinated with a pre-emptive strike on the Soviet Union – and he communicated his willingness to escalate indirectly to the two countries.[61]

Even Bertrand Russell, a lifelong pacifist, pressed for Russia to be given an ultimatum: give up your nuclear ambitions, join a 'world government' – or face war. 'I am inclined', he said during a talk to the Royal Empire Society in 1947, 'to think that Russia would acquiesce; if not, provided this is done soon, the world might survive the resulting war and emerge with a single government such as the world needs.'[62] Like von Neumann, Russell believed the USSR to be an expansionist, totalitarian regime that, with the demise of Nazi Germany, now posed the biggest threat to world peace.[63]

Von Neumann appears to have lost his appetite for a pre-emptive strike when it became obvious that the Soviet Union had enough nuclear bombs to retaliate. After a conversation in 1954 with Oswald Veblen, who had come round to the idea, he wrote to Klári. 'I told him that I felt a "quick" war was academic by now,' von Neumann said, 'since it would now – or within a rather short time – hardly be quick.'[64]

Ironically, just as von Neumann was abandoning the doctrine of preventive war, it effectively became US policy. On 12 January 1954, John Foster Dulles, Eisenhower's secretary of state, announced that America could meet even the smallest military provocation with the full force of its nuclear arsenal. 'We want, for ourselves and the other free nations, a maximum deterrent at a bearable cost,' Dulles said. 'Local defense will always be important. But there is no local defense which alone will contain the mighty landpower of the Communist world. Local defenses must be reinforced by the further deterrent of massive retaliatory power.'

Dulles may have formed his idea of 'Massive Retaliation' as early as 1948. Many in RAND were appalled by the implications of the strategy, which they believed was perilously close to sanctioning a first strike.[65]

Von Neumann's early support for preventive war has ensured that he is often remembered as an unremittingly hardline cold warrior. George Kistiakowsky, who worked with von Neumann at Los Alamos, described him in 1984 as 'a hawk, unmistakable, by modern standards'.[66]

Von Neumann was, however, a complex character. He loyally served both Democratic and Republican postwar administrations – and hated the paranoiac persecution of leftist and liberal academics that was gathering pace under Senator Joseph McCarthy. At the same time as von Neumann was busily trying to convince the US to bomb the Soviet Union into submission, he was also defending his friend Robert Oppenheimer in secret hearings held by the Atomic Energy Commission (AEC) to determine whether the leader of the US atom bomb project posed a security risk. Oppenheimer was by then director of the IAS. Awkwardly for von Neumann, the chairman of the AEC was Lewis Strauss, a trustee of the IAS who was also his friend. And one of the most damning testimonies against Oppenheimer was to be given by von Neumann's fellow Budapester Edward Teller.

Von Neumann disagreed with Oppenheimer's support for the Communist Party.[67] Nonetheless, he was always a passionate champion of Oppenheimer's role in the atom bomb project – and dismissed the idea that he would ever be disloyal to his country as dangerous nonsense. 'Robert at Los Alamos was so very great,' he maintained, 'in Britain they would have made him an Earl. Then, if he walked down the street with his fly buttons undone, people would have said – look, there goes the Earl. In postwar America we say – look, his fly buttons are undone.'[68]

A particularly damaging accusation made against Oppenheimer during the trial was that he had sought to retard America's hydrogen bomb programme during his tenure as chairman of the AEC General Advisory Committee, of which von Neumann was also a member. Oppenheimer did in fact oppose it – on technical grounds – and the two men vigorously but respectfully disagreed on the matter. Teller, on the other hand, never forgave Oppenheimer for trying to nix his brainchild.

Von Neumann quickly rounded up some key witnesses: well-respected scientists who had disagreed with Oppenheimer over the Super but were still sure that he was in no way a security risk. When von Neumann's turn came to testify, he skilfully defended Oppenheimer in front of the panel. Whatever reservations Oppenheimer may have had about the hydrogen bomb, von Neumann said, he put them aside as soon as Truman announced on 31 January, 1950 that the US would develop the weapon. When prosecutors asked him if someone with close communist affiliations should be employed in a sensitive job, von Neumann replied that such connections before the war, when the Soviet threat was not obvious, were irrelevant. 'We were all little children with respect to the situation which had developed, namely, that we suddenly were dealing with something with which one could blow up the world,' he told the panel. 'None of us had been educated or conditioned to exist in this situation, and we had to make our rationalization and our code of conduct as we went along.'[69]

The four-week hearing had begun on 12 April 1954. Oppenheimer's security clearance was stripped on 29 June. In 2009, historians with access to the KGB archives found that Soviet intelligence had made many attempts to recruit Oppenheimer – but failed.[70] They concluded he was not a spy. When Eisenhower appointed von Neumann to the AEC in 1955, some of his closest friends asked how he could bring himself to join the very agency that had persecuted Oppenheimer. Veblen, who had brought von Neumann to the US in the first place, never forgave him, even refusing to visit the dying von Neumann in hospital despite letters from Klári begging him to come. Oppenheimer was more understanding. He told Klári, 'There have to be good people on both sides.'[71]

At RAND, game theory was being applied to the most pressing military problem of the time – how to avoid or survive a nuclear conflict with the Soviet Union. Though there is scant evidence von Neumann viewed international conflicts in game theoretic terms – others did. Game theory, and the Prisoner's Dilemma in particular, quickly became the analytical instrument of choice for American foreign policy in the febrile atmosphere of fear and paranoia that persisted into the late twentieth century. 'The Cold War,' says historian Paul Erickson, 'came to be seen by many as the ultimate game that game theory was meant to analyze.' Game theoretical analysis was so ubiquitous, he adds, that 'the histories of many key geopolitical events of the Cold War era . . . would be rewritten through this lens, to the point that post hoc analysis and history could become difficult to distinguish'.[72] Among the first of RAND's analysts to study the question of nuclear deterrence was Albert Wohlstetter, whose reputation for hard-nosed, fact-based analysis helped make him one of the most influential 'defense intellectuals' of the twentieth century.

From the first, Wohlstetter was an unlikely hawk. He was a

logician, who as a teenager wrote an article for the journal *Philosophy of Science* that prompted Albert Einstein to invite him to tea. The physicist, who proclaimed Wohlstetter's article to be 'the most lucid extrapolation of mathematical logic he had ever read', wanted to hash out the finer points of the piece with the seventeen-year-old.

At Columbia University, Wohlstetter joined a communist splinter group, the League for a Revolutionary Workers Party. Had the membership records of the group not been lost in a freak accident, Wohlstetter might never have made it into RAND at all. He joined the think tank in 1951 as a consultant to the maths division. His wife, Roberta, worked there too, first as a book reviewer in the Social Sciences Division, then as a highly respected analyst whose authoritative study on surprise attacks, published in 1962 as *Pearl Harbor: Warning and Decision*,[73] would still be cited years later, including in 2004 by the 9-11 Commission.

Wohlstetter soon grew bored with the humdrum methodological work he had been assigned in the Mathematics Division. He was an aesthete with a taste for fine wine and haute cuisine, who would host classical music concerts at his modernist home nestled in the Hollywood Hills. He pined for more interesting challenges. One came his way when Charles Hitch, the head of RAND's Economics Division, asked him to assess the best places to locate overseas bases for Strategic Air Command (SAC), America's fleet of nuclear bombers. Fearing that the study would be every bit as dull as the assignments he was trying to escape from, Wohlstetter initially turned Hitch down. After turning the problem over in his mind that weekend, he changed his mind.

Wohlstetter had been disgusted by the US decision to drop atom bombs on Hiroshima and Nagasaki, an act he regarded as cruel and unnecessary. Here was a chance to reshape American nuclear strategy in a way that might spare cities in a future conflagration. He had detected a simple but serious conundrum at the heart of the basing question that piqued his interest: if your bases were sited close to the enemy, the enemy would also be close to you. It was not the first time someone had noticed this dilemma, but two intellectual influences led Wohlstetter to give the issue more serious thought: the first was game theory, the second was his wife.

Not only was game theory impossible to avoid in the Mathematics

Division by 1951 but J. C. C. McKinsey, a good friend from his Columbia days who was also employed by RAND, was busy writing on the subject. Wohlstetter was not so interested in the mathematical intricacies of the theory but he did note the central premise: that in formulating a strategy, one had always to account fully for the actions of a rational enemy. His wife, Roberta, on the other hand, was investigating why the United States was completely unprepared for the Japanese attack on its ships. So Wohlstetter was acutely aware that no study of overseas American bases would be complete if it did not account for the possibility of a Soviet attack. He began digging deeply into the numbers and assembled a group of mathematically minded analysts to help. Known as 'systems analysis', their method, developed at RAND, was related to operations research but with a different emphasis. Operations research was a science of the possible: what can be achieved with the equipment and supplies available? Systems analysis, by contrast, was goal-orientated – what future weapons and strategies would be necessary to a specified mission? With its tacit commitment to considering every 'rational' eventuality, systems analysis is almost megalomaniac in its ambition.

The team considered numerous scenarios from basing the bombers in the US to the SAC's preferred option of deploying them abroad. After exhaustive study, what they found was that under Strategic Air Command's plan, the US bombers stationed in Europe were sitting ducks. A Soviet pre-emptive strike, they calculated, would wipe out nearly 85 per cent of the bomber force stationed in Europe. Worse, the near-elimination of US nuclear forces could be accomplished with just 120 bombs of 40 kilotons each – about twice that of Fat Man – leaving the Soviet Union free to invade Western Europe with impunity or hold America to ransom. The option that came out on top was to use the overseas bases to refuel the planes but not to station them permanently. Refuelling in-flight, an idea favoured by some Air Force leaders, was far too costly.

The RAND study, entitled *Selection and Use of Strategic Bases*, proved completely unpalatable to the Air Force.[74] Wohlstetter's team briefed officials on over ninety occasions but met near-uniform intransigence. 'I hope none of you are taken in by all this slide-rule razzmatazz,' one colonel scoffed after hearing their presentation. A major obstacle proved to be SAC's chief, the cigar-chomping Curtis LeMay, who would serve as a model for a couple of the bellicose generals in *Dr Strangelove*.

Even more uncompromising than Arnold, LeMay had led the Twentieth Air Force's campaign of carpet-bombing Japanese cities. 'All war is immoral,' he once declared. 'If you let that bother you, you're not a good soldier.' LeMay's preferred nuclear strategy, the 'Sunday Punch', was 'Massive Retaliation' by another name: a no-holds-barred attack on the Soviet Union with every atom bomb at SAC's disposal in response to any aggression. If there was, as RAND now claimed, a risk of a Soviet surprise attack then, why, that was sufficient reason to hit *them* first. It was not so much a war plan, Kahn told SAC officers, as a 'war orgasm'.

Eventually, RAND engineered a meeting with General Thomas White, then acting Air Force chief of staff. A worried Wohlstetter, convinced that he had decisively demonstrated America was on the brink of a war that it would lose, presented his team's analysis one more time and convinced White of its importance. In October 1953, two months after the RAND presentation, the Air Force agreed to harden bases to atomic attack and reduce the number of aircraft stationed overseas to the bare minimum. Wohlstetter's recommendations were never fully acted upon. Instead, SAC adopted the plan the RAND team had rejected as too expensive: refuel US-based bombers in mid-air and reduce SAC's dependency on foreign bases. Still, RAND had forced a major rethink of Air Force strategy on the basis of theoretical mathematical projections.

Wohlstetter continued thinking about America's nuclear defences. At the end of the 1950s, he laid out his ideas in 'The Delicate Balance of Terror', an article that helped to shape strategic thought in America for decades.[75] Wohlstetter attacked the widely held belief that the existence of two nuclear powers eliminated the risk of an all-out global war. There was no atomic stalemate, he argued. The West had been lulled into a false sense of security by imagining that Soviet leaders favoured attacking in ways that would result in plenty of warning for the US. 'However attractive it may be for us to narrow Soviet alternatives to these, they would be low in the order of preference of any reasonable Russian planning war,' he argued and, echoing game theory's minimax principle, he added that 'In treating Soviet strategies it is important to consider Soviet rather than Western advantage and to consider the strategy of both sides quantitatively. The effectiveness of our own choices will depend on a most complex numerical interaction of Soviet and Western plans.'

His conclusion was that there could be no atomic stalemate and no let-up in vigilance. Any American vulnerability was both an invitation for the Russians to launch an immediate attack and reason for the US to pre-empt them by launching first. But as Wohlstetter was busily briefing commanders in Washington, von Neumann brought news of a weapon that threatened to make bombers obsolete altogether.

In 1950, RAND had produced a number of studies concluding that the development of long-range ballistic missiles should be an Air Force priority.[76] Partly in response, the Defense Department initiated the Atlas Missile Project in 1951 to determine whether a rocket armed with a 3,000-pound warhead could be sent to destroy cities more than 5,000 miles away. But the bombs dropped on Hiroshima and Nagasaki were many times too heavy for Atlas, and the first US thermonuclear test, codenamed 'Ivy Mike', on 1 November the previous year had used a 74 metric-ton device, too heavy to be loaded on a plane, never mind a missile. Atlas was consequently a rather lower-priority project – a moonshot for the future. But in 1953, von Neumann, accompanied by Teller, told RAND's physicists that the weaponeers at Los Alamos were on the verge of being able to make hydrogen bombs light enough to fit on rockets, and Hap Arnold's vision of city-destroying projectiles that 'come streaking out of somewhere' could quickly be realized.

The first to hear at RAND were Ernst Plesset, head of the Physics Division, and David Griggs, the former US Air Force chief scientist who had held Plesset's job in RAND's early days and was now a consultant for the think tank. They passed on what they knew about the Los Alamos work to another RAND physicist, Bruno Augenstein, who began working through the implications of the news. The first Soviet hydrogen bomb test on 12 August 1953 gave his work added urgency. The 'Mike' device had been housed in a gigantic vacuum flask to keep its liquid deuterium fuel cool. Analysis of fallout from the Soviet test had revealed the presence of lithium, an indication that lithium deuteride, a solid at room temperature, might have served as a fuel for their device. If the Soviets had managed to make lithium deuteride in quantity, they could probably make weapons small enough to be loaded onto a bomber – or perhaps even launched on a rocket.

The Atlas programme's managers had asked for missiles to be built to demanding specifications. They wanted a rocket that would fly halfway around the world at six times the speed of sound and land within a half-mile of the target. Augenstein realized that a lightweight hydrogen bomb rendered those requirements unnecessary. Using figures he got from Los Alamos, he calculated that a bomb weighing less than 1,500 pounds would produce a blast of several megatons. His research also suggested that the Russians would have trouble shooting down missiles travelling much slower than the speeds envisioned by those in charge of the project. Augenstein's most significant discovery, however, was that the destructive power of the new warhead meant that a missile that landed between 3 and 5 miles away from the target would be sufficient – and within the capabilities of contemporary missile guidance technology. The US could develop ICBMs years earlier than the Atlas programme envisaged, perhaps as soon as 1960. And Augenstein knew that if he had reached these conclusions, the Russians had too – only earlier.

Augenstein's report landed on Collbohm's desk on 11 December 1953. Much impressed, Collbohm took it to Washington the next day to try to convince senior Air Force officers of the urgency of the situation. They wanted to wait. In October, the Air Force had itself assigned eleven of the country's leading scientists and engineers to examine the feasibility of ICBMs – with von Neumann as their chairman. Codenamed the 'Teapot Committee', the panel had started their deliberations the previous month. Augenstein returned to RAND to prepare a formal report that would flesh out the technical details of the missiles and estimate how many missiles of lesser accuracy would be needed to destroy Soviet cities.

The Air Force received Augenstein's final analysis, entitled *A Revised Development Program for Ballistic Missiles of Intercontinental Range*, on 8 February, 1954 – two days ahead of von Neumann's committee. Their conclusions and recommendations were almost identical. Within a couple of months of the two reports, the US had relaxed the tight strictures imposed on the Atlas project and started a fast-track programme to develop missiles tipped with H-bombs.

On 21 August 1957, the USSR sent the R-7 Semyorka streaking nearly 4,000 miles through the air from Baikonur Cosmodrome in Kazakhstan. A few weeks later, essentially the same rocket put Sputnik into Earth orbit. The first successful flight of an Atlas rocket, a direct result of the programme that von Neumann helped to accelerate, took place on 28 November 1958. The armed Atlas D entered service in September 1959, within the timescale that Augenstein had envisaged in his report. The ICBM's threat of raining death from the skies at the lunatic push of a button has hung over the world ever since.

At RAND, the melding of von Neumann's game theory with defence policy continued apace and found a vocal advocate in the corpulent form of Herman Kahn. To the chagrin of his colleagues, Kahn toured the US cheerfully recasting their theories as provocatively as possible, rapidly becoming the most infamous of RAND's 'defence intellectuals'. 'Thinking about the Unthinkable', an idea virtually synonymous with the RAND ethos, was the title of one of Kahn's books and reflected game theory's rational, at times pathological, precept of imagining the worst possible response to any policy. A true 'jester of death',[77] Kahn played deterrence theory for laughs, delivering his lines with relentless deadpan humour while he reasoned his way through the apocalypse, always willing to go a step further than anyone else. Beneath all his shtick, Kahn was deadly serious. When his fellow strategist Bernard Brodie worried during a RAND meeting that even a nuclear strike solely on Russian military targets located outside cities would still kill two million, Kahn piped up that a strategy that left '*only*' two million dead could not be dismissed out of hand.[78]

Kahn was a physicist by training – during his early days at RAND, he had run Monte Carlo hydrogen-bomb-related simulations on the dozen or so high-speed computers then operating in the US.[79] He was one of about seventy researchers at RAND given the high-level security clearances necessary to work on bomb design and he worked in the Physics Division, separated from the rest of the building by an electronic door. But Kahn was not content to skulk behind a closed door, calculating. He ambled along the corridors of RAND, sniffing the air for stimulating problems that would help make his name. Game theory piqued his interest for a while and, inspired, he began (but never finished) a book on its application to military planning.[80] Then he found Wohlstetter and quickly realized that the new field of nuclear deterrence theory matched his unique talents. Formal game theory was too constraining for Kahn, but its assumptions often hover below the surface of his work.

Kahn eventually compiled his early lectures on deterrence into a massive tome of more than 600 pages and gave a copy to Wohlstetter, who advised him to burn it.[81] Instead, he published it, and 'On Thermonuclear War' went on to sell a remarkable 30,000 copies in hardback.[82] In it, Kahn asserted that nuclear war with the Soviet Union might be survivable and 'would not preclude normal and happy lives for the majority of survivors and their descendants'. 'Will the survivors envy the dead?' Kahn asked at the foot of one table before concluding that they would not. The table, headed 'Tragic but Distinguishable Postwar States', listed numbers of dead (from 2 to 160 million) against the time Kahn thought that the economy would take to recover (up to 100 years).

It was the sort of blithe strategizing that Stanley Kubrick, who read Kahn's book closely, would satirize so brilliantly in *Dr Strangelove*.[83] 'Now, the truth is not always a pleasant thing,' says the film's ultra-hawkish general, Buck Turgidson, as he makes the case for a massive strike on the Soviet Union,

> but it is necessary now to make a choice, to choose between two admittedly regrettable, but nevertheless distinguishable postwar environments: one where you got twenty million people killed, and the other where you got a hundred and fifty million people killed . . . Mr President, I'm not saying we wouldn't get our hair mussed. But I do say no more than ten to twenty million killed, tops.[84]

It was also the type of strategizing that sickened many of Kahn's critics. Pacifists, including Russell, felt Kahn had inadvertently made the case for universal disarmament. One of the most notorious reviews of Kahn's book, by mathematician James Newman, appeared in

Scientific American.[85] 'Is there really a Herman Kahn? It is hard to believe . . . No one could write like this; no one could think like this,' wrote Newman. 'Perhaps the whole thing is a staff hoax in bad taste.'[86] 'This is a moral tract on mass murder,' Newman continued, 'how to plan it, how to commit it, how to get away with it, how to justify it.'

Kahn was appalled by Newman's review – so appalled, in fact, that he soon started writing a sequel. The success of *On Thermonuclear War* helped Kahn win a million-dollar grant from the Rockefeller Foundation, and with it he set up his own think tank, the Hudson Institute in New York, which he called a 'high-class RAND'.

RAND's zeal for game theory research had largely fizzled out by the early 1960s, though its tenets and approach were embedded in the think tank's culture from the systems analysis it pioneered to the prescriptions of its defence policy experts. One of the last at RAND to turn game theory to the question of nuclear deterrence was Harvard University economist Thomas Schelling. He conceived of war as bargaining by other means and he set out his new approach to conflict in a 1958 paper.[87] 'On the strategy of pure conflict – the zero-sum games – game theory has yielded important insight and advice,' he says. 'But on the strategy of action where conflict is mixed with mutual dependence – the non-zero-sum games involved in wars and threats of war, strikes, negotiations, criminal deterrence, class war, race war, price war, and blackmail; maneuvering in a bureaucracy or a social hierarchy or in a traffic jam; and the coercion of one's own children – traditional game theory has not yielded comparable insight or advice.' Much of Schelling's career would be spent addressing this lacuna.

Schelling also demonstrated that even in many situations where no explicit communication was allowed or possible, players were able to coordinate their responses for mutual benefit far more often than game theory predicted. In one class experiment, Schelling asked his students to suppose they had to meet a stranger in New York City the next day but had no means of communicating with them. When and where should you meet? The most common answer was Grand Central Station at noon. Schelling called these unanticipated solutions to cooperative games 'focal points', and they showed the limits of theory. 'One cannot,' Schelling concluded, 'without empirical evidence, deduce what understandings can be perceived in a non-zero sum game of maneuver any more than one can prove, by purely formal deduction, that a particular joke is bound to be funny.'[88] But Schelling warned such tacit communication may not be enough to prevent a conflict escalating into a nuclear exchange.[89] He recommended strengthening channels between the leaders of the US and Soviet Union several years before the Cuban Missile Crisis would expose how poor communication between the two might have disastrous consequences. Like his colleagues at RAND, Schelling thought the threat of 'massive retaliation' was not credible when the other side had the capacity to strike back equally massively. The Soviet Union would never be deterred by a policy that would mean national suicide for the US if it was ever acted upon.

Von Neumann's last word on nuclear deterrence was published in 1955. *Defense in Atomic War* expresses the new bomb's power in stark terms.[90] 'The increases of firepower which are now before us are considerably greater than any that have occurred before,' he says.

> The entire tonnage of TNT dropped on all battlefields during all of World War II by all belligerents was a few million tons. We delivered more explosive power than this in a single atomic blast.[91] Consequently, we can pack in one airplane more firepower than the combined fleets of all the combatants during World War II.

Von Neumann explains that while historically the upper hand in a war might oscillate between combatants, this step-change in the destructiveness of weapons now available to the superpowers had changed the nature of war completely. 'The difficulty with atomic weapons and especially missile-carried atomic weapons,' he argues, 'will be that they can decide a war, and do a good deal more in terms of destruction in less than a month or two weeks. Consequently, the nature of technical surprise will be different from what it was before.' As a full-force nuclear attack from one side is impossible to defend against, 'this will probably mean you will be forced not to 'do your worst' at all times, because then when the enemy does his worst you cannot defend against it . . . Hence, you may have to hold this trump card in reserve.'

A strategy of exercising restraint – rather than the reflexive 'wargasm' of massive retaliation – was by 1960 a popular position at RAND.[92] The patently hollow threat of going nuclear in response to the smallest attack with conventional forces was doing nothing to discourage Soviet aggression. Even as Dulles made his 1954 'massive retaliation' speech, the US was becoming increasingly ensnared by a conventional conflict in Vietnam, and by the end of the same year the secretary of state himself was questioning the policy. In a letter to President Eisenhower in December, Dulles asked whether the US was 'prepared to deal adequately with the possible "little wars" which might call for punishment related to the degree and locality of the offense, but which would not justify a massive retaliation against the Soviet Union itself'.

RAND's response to the problem of fighting 'little wars' was the doctrine of 'counterforce'. Pioneered by RAND's Brodie, beefed up with game theory matrices by RAND analysts, then articulated most comprehensively by William Kaufmann, counterforce proposed sparing cities in the first instance. In response to Soviet aggression, the US would fire a small number of weapons at non-urban military targets, then use the threat of a well-protected nuclear reserve force as a bargaining lever to halt further escalation. Kaufmann hoped that if the Russians *did* retaliate, they too would avoid hitting cities. Civilian lives could be spared, and with time for negotiations to defuse the crisis perhaps an all-out nuclear exchange could be averted.

Counterforce was quintessential RAND, epitomizing the think tank's quest for, as Kahn put it, 'more reasonable forms of using violence'. The problem was that avoiding bloodshed was not a universally popular idea within the US military. The Strategic Air Command (SAC), in charge of America's bombers and ICBMs, was particularly hostile to the new strategy.

Counterforce would find more receptive ears in government after the 1960 election of President John F. Kennedy, whose campaign was covertly aided by some of RAND's experts. Among them was Daniel Ellsberg, who would leak the top secret 'Pentagon Papers', with their damaging revelations about the Vietnam War, to the press in 1971. Kennedy's defence secretary, Robert McNamara, would hustle a host of RAND analysts, including Kaufmann, into the White House. The young 'Whizz Kids', as they became known, earned the enmity of the Air Force, their one-time sponsors, as their systems analysis studies undermined prized bomber and rocket projects – but supported accelerating the Navy's submarine-launched Polaris missiles and expanding the Army's conventional forces. Tired of having their defeats rubbed in their faces by the new gang of Ivy Leaguers, the Air Force soon hired their own analysts, and the Navy and the Army followed suit. RAND's methods were embedded in US military thinking, shaping the country's approach to theoretical nuclear conflicts – and to the very real 'little wars' to come in Southeast Asia and elsewhere.

In June 2019, the Pentagon accidentally published to its website the US military's guidelines for planning and executing small-scale nuclear warfare. The sixty-page document, JP 3-72 on *Joint Nuclear Operations*, was quickly removed – but not before it had been downloaded by the Federation of American Scientists (FAS), a charity founded by Manhattan Project researchers in 1945 that is devoted to peaceful uses of atomic energy.[93] The report's focus is worst-case scenarios, and its emphasis is on fighting wars, rather than deterrence. Critics aver that such talk of limited nuclear war helps convince America's enemies that the US would be prepared to use the bomb – increasing the chances that *someone* will. By imagining the worst, the worst is brought a step closer to being realized.

The dilemma is one that RAND's analysts would have recognized. Seventy years after they began applying the tools of game theory to nuclear strategy, the stakes are higher than ever, the bombs dropped on Hiroshima and Nagasaki mere firecrackers compared to some of the bombs in the American and Russian arsenals. More countries now possess the weapons, and others are threatening to produce them. The expertise required to build a device is now widespread enough that it is not unfeasible that a well-organized terrorist group could build one. So a strategy document from the world's most powerful bearer of nuclear arms might be expected to be unrecognizable in its scope compared with the stuff produced by the Cold-War-hardened strategists of the 1950s. Much of the report, however, is strikingly familiar – not least the epigraph that begins the third chapter on 'Planning and Targeting': 'My guess is that nuclear weapons will be used sometime in the next hundred years, but that their use is much more likely to be small and limited than widespread and unconstrained.' It comes from the 1962 book *Thinking About the Unthinkable*. Its source is Herman Kahn.

'The most spectacular event of the past half century is one that did not occur,' said Schelling in 2005, a couple of days before collecting his Nobel Prize. 'We have enjoyed sixty years without nuclear weapons exploded in anger.' Schelling attributed our 'stunning good fortune' to an unspoken taboo against the use of even the 'smallest' bomb. Should the horror of Hiroshima and Nagasaki fade from the public consciousness, he warned, should more nations or even terrorist groups acquire nuclear weapons, there is no guarantee they will share this 'nearly universal revulsion' against their use. We are on borrowed time.

By the mid-1950s, von Neumann too was on borrowed time. Perhaps it was all the nuclear weapons tests he had attended, or his unhealthy diet or simply bad luck, but cancer was slowly metastasizing through his body. Unaware of the illness, he spent his last years in frenzied activity. While mathematicians tend to produce their best work in their twenties and view middle age as the twilight of their productive years, von Neumann would produce his most original work yet. He began a quest to understand the phenomenal powers of sophisticated machines from the ground up; in particular the high-speed computers he had helped build and the most complex and mysterious of them all, the human brain.

8

The Rise of the Replicators

Machines to make machines and machines to make minds

'"The androids," she said, "are lonely too."'

Do Androids Dream of Electric Sheep?,

Philip K. Dick, 1968

The odd angles, harsh lighting and wobbly movements of the handcam give the video of a 3D printer at work the air of an amateur porn film. In an odd sort of way it is, for the video shows a machine caught, *in flagrante*, reproducing. The 'Snappy' is a RepRap – a self-copying 3D printer – that is able to print around 80 per cent of its own parts.[1] If a printer part breaks, you can replace it with a spare that you printed earlier. For the cost of materials – the plastic filament that is melted and extruded by a printer costs $20–50 a kilo – you can print most of another printer for a friend. Just a handful of generic metal parts such as bolts, screws and motors and the electronic components must be bought to complete assembly.

Engineer and mathematician Adrian Bowyer first conceived the idea he calls 'Darwinian Marxism' in 2004 – that eventually everyone's home will be a factory, producing anything they want (as long as it can be made out of plastic, anyway). Engineers at Carleton University in Ottawa are working on filling in that stubborn last 20 per cent, to create a printer that can fully replicate itself even if you don't have a DIY shop handy. Specifically, they are thinking about using only materials that can be found on the surface of the moon. Using a RepRap as their starting point, the researchers have begun to design a rover that will print all its parts and the tools it needs to copy itself using only raw materials harvested in situ by, for example, smelting lunar rock in a solar furnace.[2] They have also made experimental motors and computers with McCulloch–Pitts-style artificial neurons to allow their rover to navigate. Semiconductor-based electronic devices would be practically impossible to make on the moon, so in a charming 1950s twist, they plan to use vacuum tubes instead. 'When I came across RepRap, although it was a modest start, for me it was catalytic,' says Alex Ellery, who leads the group. 'What began as a side project now consumes my thoughts.'[3]

Once they are established on the moon, Ellery's machines could multiply to form a self-expanding, semi-autonomous space factory making . . . virtually anything. They might, for example, print bases ready for human colonizers – or even, as Ellery hopes, help to mitigate global warming by making swarms of miniature satellites that can shield us from solar radiation or beam energy down to Earth.[4]

The inspiration for all these efforts and more is a book entitled *Theory of Self-reproducing Automata*; its author, John von Neumann. In between building his computer at the IAS and consulting for the government and industry, von Neumann had begun comparing biological machines to synthetic ones. Perhaps, he thought, whatever he learned could help to overcome the limitations of the computers he was helping to design. 'Natural organisms are, as a rule, much more complicated and subtle, and therefore much less well understood in detail, than are artificial automata,' he noted. 'Nevertheless, some regularities which we observe in the organization of the former may be quite instructive in our thinking and planning of the latter.'[5]

After reading the McCulloch and Pitts paper describing artificial neural networks, von Neumann had grown interested in the biological sciences and corresponded with several scientists who were helping to illuminate the molecular basis of life, including Sol Spiegelman and Max Delbrück. In his usual way, von Neumann dabbled brilliantly, widely and rather inconclusively in the subject but intuitively hit upon a number of ideas that would prove to be fertile areas of research for others. He lectured on the mechanics of the separation of chromosomes that occurs during cell division. He wrote to Wiener proposing an ambitious programme of research on bacteriophages, conjecturing that these viruses, which infect bacteria, might be simple enough to be studied productively and large enough to be imaged with an electron microscope. A group of researchers convened by Delbrück and Salvador Luria would over the next two decades shed light on DNA replication and the nature of the genetic code by doing exactly that – and produce some of the first detailed pictures of the viruses.

One of von Neumann's most intriguing proposals was in the field of protein structure. Most genes encode proteins, molecules which carry out nearly all important tasks in cells, and are the key building block of muscles, nails and hair. Though proteins were being studied extensively in the 1940s, no one knew what they looked like, and the technique that was eventually used to study them, X-ray crystallography, was still in its infancy. The shape of a protein can be deduced from the pattern of spots produced by firing X-rays at crystals of it. The problem was, as von Neumann quickly realized, the calculations that would need to be done to reconstruct the protein's shape far exceeded the capacity of the computers available at the time. In 1946 and 1947, he met with renowned American chemist Irving Langmuir and the brilliant mathematician Dorothy Wrinch, who had developed the 'cyclol' model of protein structure. Wrinch pictured proteins as interlinked rings, a conjecture that would turn out to be wrong, but von Neumann thought that scaling up the experiment some hundred-million-fold might shed light on the problem. He proposed building centimetre-scale models using metal balls and swapping X-rays for radar technology, then comparing the resulting patterns with those from real proteins. The proposal was never funded but demonstrates how far von Neumann's interests ranged across virtually every area of cutting-edge science. In the event, advances in practical techniques as well as theoretical ones – and a great deal of patience – would be required before X-ray crystallography began to reveal protein structures – as it would, starting in 1958.

From 1944, meetings instigated by Norbert Wiener helped to focus von Neumann's thinking about brains and computers. In gatherings of the short-lived 'Teleological Society', and later in the 'Conferences on Cybernetics', von Neumann was at the heart of discussions on how the brain or computing machines generate 'purposive behaviour'. Busy with so many other things, he would whizz in, lecture for an hour or two on the links between information and entropy or circuits for logical reasoning, then whizz off again – leaving the bewildered attendees to discuss the implications of whatever he had said for the rest of the afternoon. Listening to von Neumann talk about the logic of neuro-anatomy, one scientist declared, was like 'hanging on to the tail of a kite'.[6] Wiener, for his part, had the discomfiting habit of falling asleep during discussions and snoring loudly, only to wake with some pertinent comment demonstrating he had somehow been listening after all.

By late 1946, von Neumann was growing frustrated with the abstract model of the neuron that had informed his EDVAC report on programmable computers a year earlier. 'After the great positive contribution of Turing-cum-Pitts-and-McCulloch is assimilated, the situation is rather worse than before,' he complained in a letter to Wiener.

> Indeed, these authors have demonstrated in absolute and hopeless generality, that ... even one, definite mechanism can be 'universal'. Inverting the argument: Nothing that we may know or learn about the functioning of the organism can give, without 'microscopic', cytological work any clues regarding the further details of the neural mechanism ... I think you will feel with me the type of frustration that I am trying to express.

Von Neumann thought that pure logic had run its course in the realm of brain circuitry and he, who had set a record bill for broken glassware while studying at the ETH, was sceptical that fiddly, messy lab work had much to offer. 'To understand the brain with neurological methods,' he says dismissively,

> seems to me about as hopeful as to want to understand the ENIAC with no instrument at one's disposal that is smaller than about 2 feet across its critical organs, with no methods of intervention more delicate than playing with a fire hose (although one might fill it with kerosene or nitroglycerine instead of water) or dropping cobblestones into the circuit.

To avoid what he saw as a dead end, von Neumann suggested the study of automata be split into two parts: one devoted to studying the basic elements composing automata (neurons, vacuum tubes), the other to their organization. The former area was the province of physiology in the case of neurons, and electrical engineering for vacuum tubes. The latter would treat these elements as idealized 'black boxes' that would be assumed to behave in predictable ways. This would be the province of von Neumann's automata theory.

The theory of automata was first unveiled in a lecture in Pasadena on 24 September 1948, at the Hixon Symposium on Cerebral Mechanisms in Behaviour and published in 1951.[7] Von Neumann had been thinking about the core ideas for some time, presenting them first in informal lectures in Princeton two years earlier. His focus had shifted subtly. Towards the end of his lecture, von Neumann raises the question of whether an automaton can make another one that is as complicated as itself. He notes that at first sight, this seems untenable because the parent must contain a complete description of the new machine *and* all the apparatus to assemble it. Although this argument 'has some indefinite plausibility to it,' von Neumann says, 'it is in clear contradiction with the most obvious things that go on in nature. Organisms reproduce themselves, that is, they produce new organisms with no decrease in complexity. In addition, there are long periods of evolution during which the complexity is even increasing.' Any theory that claims to encompass the workings of artificial and natural automata must explain how man-made machines might reproduce – and evolve. Three hundred years earlier, when the philosopher René Descartes declared 'the body to be nothing but a machine' his student, the twenty-three-year-old Queen Christina of Sweden, is said to have challenged him: 'I never saw my clock making babies'.[8] Von Neumann was not the first person to ask the question 'can machines reproduce?' but he would be the first to answer it.

At the heart of von Neumann's theory is the Universal Turing machine. Furnished with a description of any other Turing machine and a list of instructions, the universal machine can imitate it. Von Neumann begins by considering what a Turing-machine-like automaton would need to make copies of itself, rather than just compute. He argues that three things are necessary and sufficient. First, the machine requires a set of instructions that describe how to build another like it – like Turing's paper tape but made of the same 'stuff' as the machine itself. Second, the machine must have a construction unit that can build a new automaton by executing these instructions. Finally, the machine needs a unit that is able to create a copy of the instructions and insert them into the new machine.

The machine he has described 'has some further attractive sides', notes von Neumann, with characteristic understatement. 'It is quite clear' that the instructions are 'roughly effecting the functions of a gene. It is also clear that the copying mechanism . . . performs the fundamental act of reproduction, the duplication of the genetic material, which is clearly the fundamental operation in the multiplication of living cells.' Completing the analogy, von Neumann notes that small alterations to the instructions 'can exhibit certain typical traits which appear in connection with mutation, lethally as a rule, but with a possibility of continuing reproduction with a modification of traits'.[9] Five years before the discovery of the structure of DNA in 1953, and long before scientists understood cell replication in detail, von Neumann had laid out the theoretical underpinnings of molecular biology by identifying the essential steps required for an entity to make a copy of itself. Remarkably, von Neumann also correctly surmised the limits of his analogy: genes do not contain step-by-step assembly instructions but 'only general pointers, general cues' – the rest, we now know, is furnished by the gene's cellular environment.

These were not straightforward conclusions to reach in 1948. Erwin Schrödinger, by now safely ensconced at the Dublin Institute for Advanced Studies, got it wrong. In his 1944 book *What Is Life?*,[10] which inspired, among others, James Watson and Francis Crick to pursue their studies of DNA, Schrödinger explored how complex living systems could arise – in seeming contradiction with the laws of physics. Schrödinger's answer – that chromosomes contain some kind of 'code-script' – failed to distinguish between the code and the means to copy and implement the code.[11] 'The chromosome structures are at the same time instrumental in bringing about the development they foreshadow,' he says erroneously. 'They are law-code and executive power – or, to use another simile, they are architect's plan and builder's craft all in one.'

Few biologists appear to have read von Neumann's lecture or understood its implications if they did – surprising, considering many of the scientists von Neumann met and wrote to during the 1940s were physicists with an interest in biology. Like Delbrück, many would soon help elucidate the molecular basis of life. That von Neumann was wary of journalists sensationalizing his work and too busy to popularize the theory himself probably did not help (Schrödinger, by contrast, had the lay reader in mind as his audience).

One biologist who did read von Neumann's words soon after their publication in 1951 was Sydney Brenner. He was later part of Delbrück's coterie of bacteriophage researchers and worked with Crick and others in the 1960s to crack the genetic code. Brenner says that when he saw the double helix model of DNA in April 1953, everything clicked into place. 'What von Neumann shows is that you have to have a mechanism not only of copying the machine but of copying the information that specifies the machine,' he explains. 'Von Neumann essentially tells you how it's done ... DNA is just one of the implementations of this.'[12] Or, as Freeman Dyson put it, 'So far as we know, the basic design of every microorganism larger than a virus is precisely as von Neumann said it should be.'[13]

In lectures over the next few years and an unfinished manuscript, von Neumann began to detail his theory about automata – including a vision of what his self-reproducing machines might look like. His work would be meticulously edited and completed by Arthur Burks, the mathematician and electrical engineer who had worked on both the ENIAC and von Neumann's computer project. The resulting book, *Theory of Self-reproducing Automata*, would only appear in 1966.[14] The first two accounts of von Neumann's ideas in print would appear in 1955, though neither was the work of automata theory's progenitor. The first of these was an article in *Scientific American*[15] by computer scientist John Kemeny, who later developed the BASIC programming language (and was Einstein's assistant at the IAS when Nash called on him). The second, which appeared in *Galaxy Science Fiction* later that year, was by the author Philip K. Dick, whose work would form the basis of films such as *Blade Runner* (1982), *Total Recall* (1990 and 2012) and *Minority Report* (2002).[16] His 'Autofac' is the tale of automatic factories set on consuming the Earth's resources to make products that no one needs – and more copies of themselves. Dick closely followed von Neumann's career, and his

Humans rise up against ceaselessly productive machines in Philip K. Dick's 'Autofac'.

story had been written the year before the *Scientific American* piece about automata appeared.[17]

Kemeny's sober write-up, entitled 'Man Viewed as a Machine', is barely less sensational than the sci-fi story. 'Are we, as rational beings, basically different from universal Turing machines?' he asks. 'The usual answer is that whatever else machines can do, it still takes a man to build the machine. Who would dare to say that a machine can reproduce itself and make other machines?' Kemeny continues, sounding deliciously like a 1950s B movie voice-over. 'Von Neumann would. As a matter of fact, he has blue-printed just such a machine.'

The first artificial creatures von Neumann imagined float on a sea of 'parts' or 'organs'. There are eight different types of part, with four devoted to performing logical operations. Of these, one organ is a signal-generator, while the other three are devoted to signal-processing.

The 'stimulus' organ fires whenever one of its two inputs receives a signal, a 'coincidence' organ fires only when both inputs do, and a fourth 'inhibitory' organ fires only if one input is active and the other is not. With these parts, the automaton can execute any conceivable set of computations.

The remaining parts are structural. There are struts, and two organs that can cut or fuse struts into larger arrangements. Lastly there is muscle, which contracts when stimulated and can, for instance, bring two struts together for welding or grasping.[18] Von Neumann suggests that the automaton might identify parts through a series of tests. If the component it has caught between its pincers is transmitting a regular pulse, for example, then it is a signal-transmitter; if it contracts when stimulated, it is muscle.

The binary instruction tape that the automaton will follow – its DNA – is encoded, ingeniously, using the struts themselves. A long series of struts joined into a saw-tooth shape forms the 'backbone'. Each vertex can then have a strut attached to represent a 'one' or be left free for a 'zero'. To write or erase the tape, the automaton adds or removes these side-chains.

From the parts that von Neumann describes it is possible to assemble a floating autofac, fully equipped to grab from the surrounding sea whatever parts it needs to make more, fertile autofacs – complete with their own physical copy of 'strut DNA'. A strut broken or added

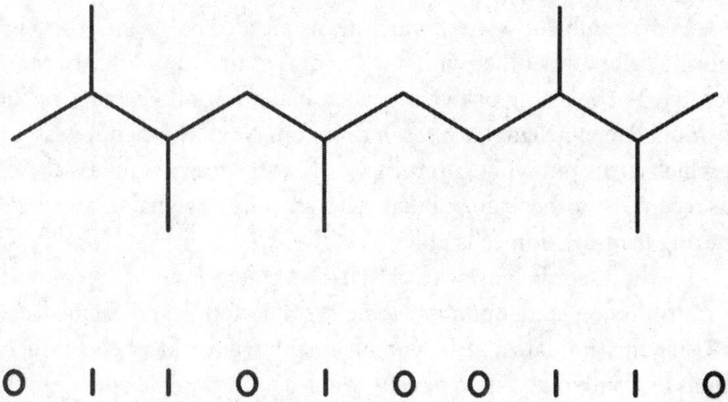

0 1 1 0 1 0 0 1 1 0

A binary tape made from struts.

by mistake is a mutation – often lethal, sometimes harmless and occasionally beneficial.

Von Neumann was not satisfied with this early 'kinematic' model of self-replication. The supposedly elementary parts from which the automaton is made are actually quite complex. He wondered if simpler mathematical organisms might spawn themselves. Did self-reproduction require three dimensions, or would a two-dimensional plane suffice? He spoke to Ulam, who had been toying with the idea of automata confined to a two-dimensional crystalline lattice, interacting with their neighbours in accordance with a simple set of rules. Inspired, von Neumann developed what would become known as his 'cellular' model of automata.

Von Neumann's self-reproducing automata live on an endless two-dimensional grid. Each square or 'cell' on this sheet can be in one of twenty-nine different 'states', and each square can only communicate with its four contiguous neighbours. This is how it works.

– Most squares begin in a dormant state but can be brought to life – and later 'killed'– by the appropriate stimulus from a neighbour.

– There are eight states for transmitting a stimulus – each can be 'on' or 'off', receive inputs from three sides and output the signal to the other.

– A further eight 'special' transmission states transmit signals that 'kill' ordinary states, sending them back into the dormant state. These special states are in turn reduced to dormancy by a signal from an ordinary transmission state.

– Four 'confluent' states, which transmit a signal from any neighbouring cell to any other with a delay of two time units. These can be pictured as consisting of two switches, *a* and *b*. Both start in the 'off' position. When a signal arrives, *a* turns on. Next, the signal passes to *b*, which turns on, while *a* returns to the 'off' position unless the cell has received another signal. Finally, *b* transmits its signal to any neighbouring transmission cells able to receive it.

– Lastly, a set of eight 'sensitized' states can morph into a confluent or transmission state upon receiving certain input signal sequences.

Using this animate matrix, von Neumann will build a behemoth. He begins by combining different cells to form more complex devices. One such 'basic organ' is a pulser, which produces a pre-set stream of pulses in response to a stimulus. Another organ, a decoder, recognizes a specific binary sequence and outputs a signal in response. Other organs were necessary because of the limitations imposed by working in two dimensions. In the real world, if two wires cross, one can safely pass over the other. Not so in the world of von Neumann's cellular automaton, so he builds a crossing organ to route two signals past each other.

Then von Neumann assembles parts to fulfil the three roles that he has identified as being vital for self-reproduction. He constructs a tape from 'dormant' cells (representing '0') and transmission cells (representing '1').[19] Adding a control unit that can read from and write in tape cells, von Neumann is able to reproduce a universal Turing machine in two dimensions. He then designs a constructing arm which snakes out to any cell on the grid, stimulates it into the desired state then withdraws.

At this point, von Neumann's virtual creature slipped its leash and overcame even his prodigious multitasking abilities. Starting in September 1952, he had spent more than twelve months on the manuscript, which was by now far longer than he had imagined would be necessary to finish the task. According to Klára, he intended to return to his automata work after discharging various pressing government advisory roles – but he never would. Ill health would overtake him within sight of his goal, and the extraordinary edifice would only be completed years after his death by Burks, who carefully followed von Neumann's notes to piece together the whole beast.

The complete machine fits in an 80x400-cell box but has an enormous tail, 150,000 squares long, containing the instructions the constructor needs to clone itself. Start the clock, and step by step the monster starts work, reading and executing each tape command to produce a carbon copy of itself some distance away. The tentacular constructing arm grows until it reaches a pre-defined point in the matrix, where it deposits row upon row of cells to form its offspring. Like the RepRaps it would one day inspire, von Neumann's automaton builds by depositing each layer on the next until assembly is finished whereupon the arm retracts back to the mother, and its progeny is free to begin its own reproductive cycle. Left to its own devices, replication continues indefinitely, and a growing line of automata tracks slowly across the vastness of cellular space into infinity.

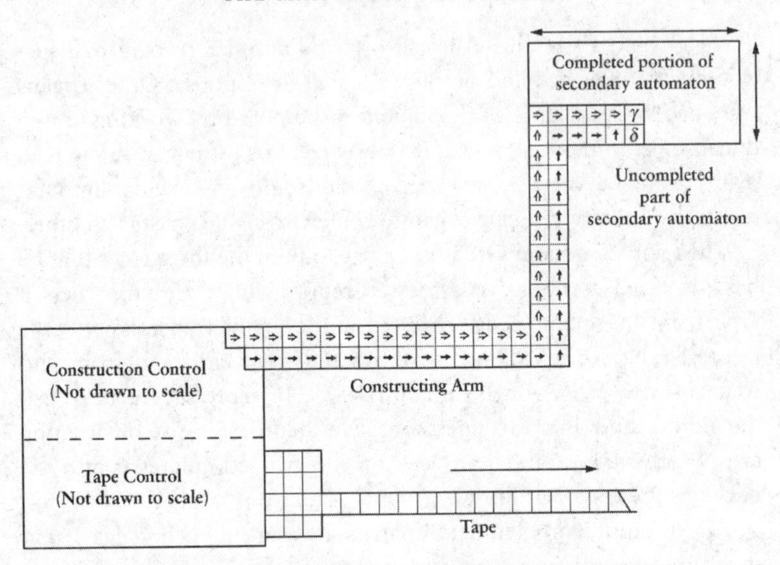

Von Neumann's self-reproducing universal constructor.

Or at least that is what Burks hoped would happen. When *Theory of Self-reproducing Automata* was published in 1966, there was no computer powerful enough to run von Neumann's 29-state scheme, so he could not be sure that the automaton he helped design would replicate without a hitch. The first simulation of von Neumann's machine, in 1994, was so slow that by the time the author's paper appeared the following year, their automaton had yet to replicate.[20] On a modern laptop, it takes minutes. With a little programming know-how, anyone can watch the multiplying critters von Neumann willed into being by the force of pure logic over half a century ago. They are also witnessing in action the very first computer virus ever designed, and a milestone in theoretical computer science.

Von Neumann's self-reproducing automata would usher in a new branch of mathematics and, in time, a new branch of science devoted to artificial life.[21] One thing stood in the way of the revolution that was to come – the first man-made replicator was over-designed. One could, with a small leap of faith, imagine the cellular automaton evolving. One could not, however, imagine such a complex being ever emerging from some sort of digital primordial slime by a fortuitous accident. Not in a million years. And yet the organic automatons that von Neumann was aping had done exactly that. 'There exists,' von Neumann wrote, 'a critical size below which the process of synthesis is degenerative, but above which the phenomenon of synthesis, if properly arranged, can become explosive, in other words, where synthesis of automata can proceed in such a manner that each automaton will produce other automata which are more complex and of higher potentialities than itself.'[22] Could a simpler automaton – ideally much simpler – ever be shown to generate, over time, something so complex and vigorous that only the most committed vitalist would hesitate to call it 'life'? Soon after *Theory of Self-reproducing Automata* appeared, a mathematician in Cambridge leafed through the book and became obsessed with exactly that question. The fruit of his obsession would be the most famous cellular automaton of all time.

John Horton Conway loved to play games and, in his thirties, managed to secure a salaried position that allowed him to do just that.[23] His early years in academia, a phase of his life he called 'The Black Blank', were stubbornly devoid of greatness. He had achieved so little he worried privately he would lose his lectureship.

That all changed in 1968, when Conway began publishing a series of remarkable breakthroughs in the field of group theory, findings attained by contemplating symmetries in higher dimensions. To glimpse Conway's achievement, imagine arranging circles in two dimensions so that there is as little space between them as possible. This configuration is known as 'hexagonal packing'; lines joining the centres of the circles will form a honeycomb-like network of hexagons. There are exactly twelve different rotations or reflections that map the pattern of circles back onto itself – so it is said to have twelve symmetries. What Conway discovered was a sought-after, rare group of symmetries for packing spheres in twenty-four dimensions. 'Before, everything I touched turned to nothing,' Conway said of this moment. 'Now I was Midas, and everything I touched turned to gold.'[24]

'Floccinaucinihilipilification', meaning 'the habit of regarding something as worthless', was Conway's favourite word. It was how he thought fellow mathematicians felt about his work. After discovering

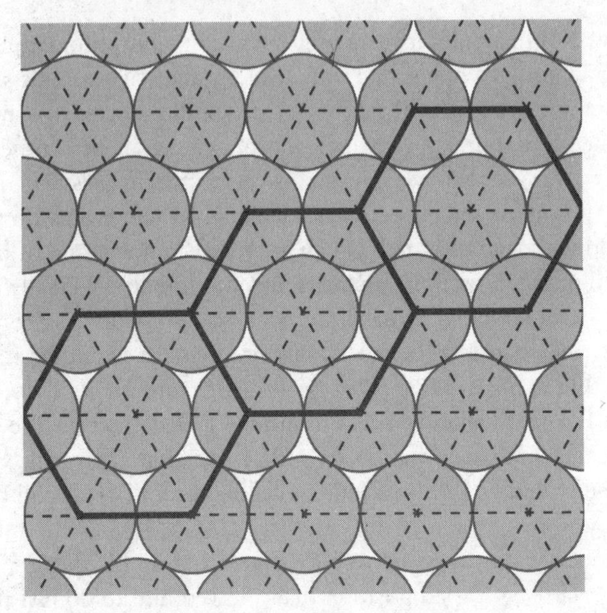

Hexagonal packing of circles.

the symmetry group that was soon named after him, as tradition decreed, Conway was free to indulge his whims with abandon. One of those whims was making up mathematical games.

With a gaggle of graduate students, Conway was often found hunched over a Go grid in the common room of the mathematics department, where they placed and removed black and white counters according to rules he and his entourage invented. Von Neumann's scheme called for twenty-nine different states. 'It seemed awfully complicated,' Conway says. 'What turns me on are things with a wonderful simplicity.'[25] He whittled down the number to just two – each cell could either be dead or alive. But whereas the state of a cell in von Neumann's scheme was determined by its own state and that of its four nearest neighbours, Conway's cell communicated with all eight of its neighbours, including the four touching its corners.

Each week, Conway and his gang tinkered with the rules that determined whether cells would be born, survive or perish. It was a tricky balancing act: 'if the death rule was even slightly too strong, then almost every configuration would die off. And conversely, if the birth rule was even a little bit stronger than the death rule, almost every configuration would grow explosively.'[26]

The games went on for months. One board perched on a coffee table could not contain the action, so more boards were added until the game crept slowly across the carpet and threatened to colonize the whole common room. Professors and students would be on their hands and knees, placing or removing stones during coffee breaks, and coffee breaks would sometimes last all day. After two years of this, the players hit a sweet spot. Three simple rules created neither bleak waste nor boiling chaos, but interesting, unpredictable variety. They were:

1. Every cell with two or three live neighbours survives.
2. A cell with four or more live neighbours dies of overcrowding. A cell with fewer than two live neighbours dies of loneliness. (In either case, the counter is removed from the board.)
3. A 'birth' occurs when an empty cell has exactly three neighbours. (A new counter is placed in the square.)

Stones were scattered, the rules applied, and generation after generation of strange objects morphed and grew, or dwindled away to nothing. They called this game 'Life'.

The players carefully documented the behaviour of different cellular constellations. One or two live cells die after a single generation, they noted. A triplet of horizontal cells becomes three vertical ones, then flashes back to the horizontal and continues back and forth indefinitely. This is a 'blinker', one of a class of periodic patterns called 'oscillators'. An innocent-looking cluster of five cells called an

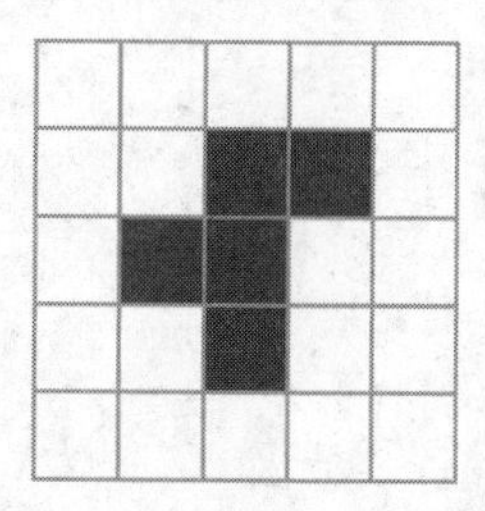

An R-pentomino from Conway's Life.

'R-pentomino' explodes into a shower of cells, a dizzying display seemingly without end. Only after months of following its progeny did they discover it stabilized into a fixed pattern some 1,103 generations later. One day in the autumn of 1969, one of Conway's colleagues called out to the other players: 'Come over here, there's a piece that's walking!' On the board was another five-celled creature that seemed to be tumbling over itself, returning to its original configuration four generations later – but displaced one square diagonally downwards. Unless disturbed, the 'glider', as it was dubbed, would continue along its trajectory for ever. Conway's automata had demonstrated one characteristic of life absent from von Neumann's models: locomotion.

Conway assembled a chart of his discoveries and sent them to his friend, Martin Gardner, whose legendary 'Mathematical Games' column in *Scientific American* had acquired a cult following among puzzlers, programmers, sceptics and the mathematically inclined. When Gardner's column about Life appeared in October 1970, letters

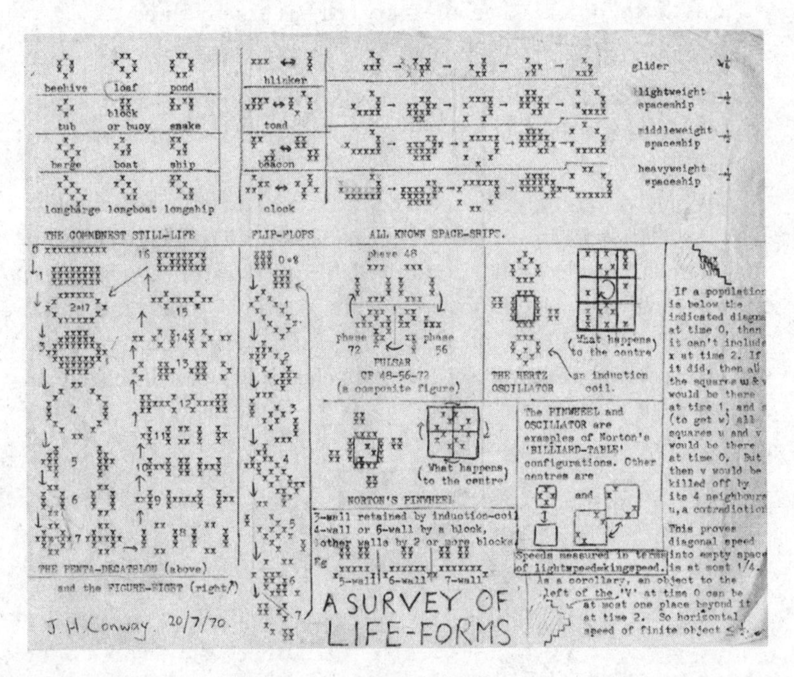

Conway's survey of life forms, as sent to Gardner.

began to pour into the mail room, some from as far afield as Moscow, New Delhi and Tokyo. Even Ulam wrote in to the magazine, later sending Conway some of his own papers on automata. *Time* wrote up the game. Gardner's column on Life became his most popular ever. 'All over the world mathematicians with computers were writing Life programs,' he recalled.

> I heard about one mathematician who worked for a large corporation. He had a button concealed under his desk. If he was exploring Life, and someone from management entered the room, he would press the button and the machine would go back to working on some problem related to the company![27]

Conway had smuggled a trick question into Gardner's column to tantalize his readers. The glider was the first piece he needed to build a Universal Turing machine within Life. Like von Neumann's immensely more complicated cellular automaton, Conway wanted to prove that Life would have the power to compute anything. The glider could carry a signal from A to B. Missing, however, was some way of producing a stream of them – a pulse generator. So Conway conjectured that there was no configuration in Life that would produce new cells ad infinitum and asked Gardner's readers to prove him wrong – by discovering one. There was a $50 prize on offer for the first winning entry, and Conway even described what the faithful might find: a 'gun' that repeatedly shoots out gliders or other moving objects or a 'puffer train', a moving pattern that leaves behind a trail. The glider gun was found within a month of the article appearing. William Gosper, a hacker at MIT's Artificial Intelligence Laboratory, quickly wrote a program to run Life on one of the institute's powerful computers. Soon, Gosper's group found a puffer train too. 'We eventually got puffer trains that emitted gliders which collided together to make glider guns which then emitted gliders but in a quadratically increasing number,' Gosper marvelled. 'We totally filled the space with gliders.'[28]

Conway had his signal generator. He quickly designed Life organisms capable of executing the basic logical operations and storing data.[29] Conway did not bother to finish the job – he knew a Turing

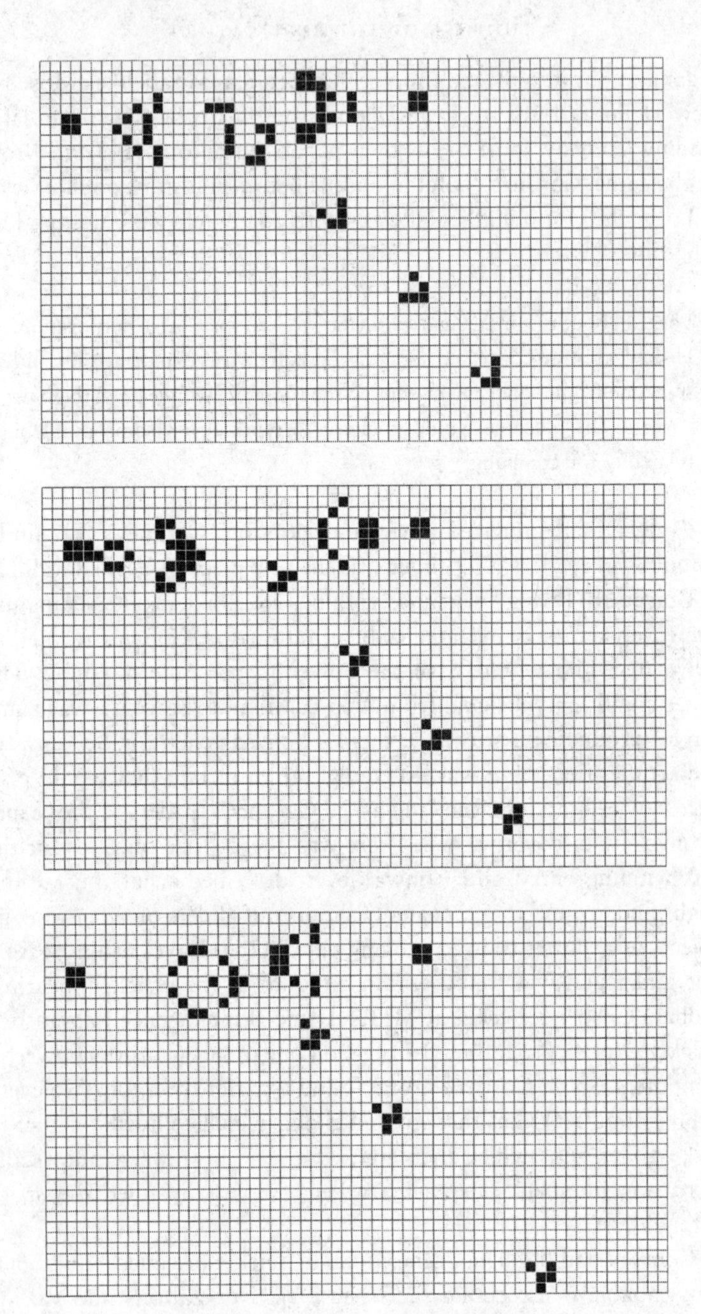

Stills of Gosper's Glider Gun.

machine could be built from the components he had assembled. He had done enough to prove that his automaton could carry out any and all computations, and that amazing complexity could arise in a system far simpler than that of von Neumann.[30] And he went a step further than even von Neumann would have dared. Conway was convinced that Life could support life itself. 'It is no doubt true that on a large enough scale Life would generate living configurations,' he asserted. 'Genuinely living. Evolving, reproducing, squabbling over territory. Writing Ph.D theses.'[31] He did, however, concede that this act of creation might need the game to be played on a board of unimaginable proportions – perhaps bigger than those of the known universe.

To Conway's everlasting chagrin, he would be remembered best as Life's inventor. Search 'Conway's Game of Life' in Google today, and flashing constellations of squares appear in one corner and begin rampaging across the screen, a testament to the game's enduring fascination.

Conway's achievements delighted automata aficionados, many of whom had been ploughing a lonely furrow. A key cluster of these isolated enthusiasts were based at the University of Michigan, where Burks had in 1956 established an interdisciplinary centre, the Logic of Computers Group, that would become a Mecca for scientists fascinated with the life-like properties of these new mathematical organisms. They shared Conway's belief that extraordinarily complex phenomena may be underpinned by very simple rules. Termites can build fabulous mounds several metres tall, but, as renowned biologist E. O. Wilson notes, 'no termite need serve as overseer with blueprint in hand'.[32]

The Michigan group would run some of the first computer simulations of automata in the 1960s, and many off-shoots of automata theory would trace their origins here.[33] One of the first in a stream of visionaries to pass through the doors of the group was Tommaso Toffoli, who started a graduate thesis on cellular automata there in 1975. Toffoli was convinced there was a deep link between automata and the physical world. 'Von Neumann himself devised cellular automata to make a reductionistic point about the plausibility of life being possible in a world with very simple primitives,' Toffoli explained. 'But even von Neumann, who was a quantum physicist, neglected completely the connections with physics – that a cellular automata could be a model of fundamental physics.'[34]

Perhaps, Toffoli conjectured, the complex laws of physics might be rewritten more simply in terms of automata. Might the strange realm of quantum physics be explained as a product of interactions between von Neumann's mathematical machines, themselves obeying just a few rules? The first step towards proving that would be to show that there existed automata that were 'reversible'. The path that the universe has taken to reach its current condition can theoretically be traced backwards (or forwards) to any other point in time (as long as we have complete knowledge of its state now). None of the cellular automata that had been invented, however, had this property. In Conway's Life, for example, there are many different configurations that end in an empty board. Starting from this blank slate, it is impossible to know which cells were occupied when the game began. Toffoli proved for his PhD thesis that reversible automata did exist and, moreover, that any automaton could be made reversible by adding an extra dimension to the field of play (so, for example, there is a three-dimensional version of Life that *is* reversible).[35]

As he cast about for a job, Toffoli was approached by Edward Fredkin, a successful tech entrepreneur who had headed MIT's pioneering artificial intelligence laboratory. Fredkin was driven by a belief that irresistibly drew him to Toffoli's work. 'Living things may be soft and squishy. But the basis of life is clearly digital,' Fredkin claimed. 'Put it another way – nothing is done by nature that can't be done by a computer. If a computer can't do it, nature can't.'[36] These were fringe views even in the 1970s, but Fredkin could afford not to care. He was a millionaire many times over thanks to a string of successful computer ventures and had even bought his own Caribbean island. Unafraid of courting controversy, Fredkin once appeared on a television show and speculated that, one day, people would wear nanobots on their heads to cut their hair. When he contacted Toffoli, Fredkin was busy assembling a group at MIT to explore his interests – particularly the idea that the visible manifestations of life, the universe and everything were all the result of a code script running on a computer. Fredkin offered Toffoli a job at his new Information Mechanics Group. Toffoli accepted.

The MIT group rapidly became the new nerve centre of automata studies. One of Toffoli's contributions to the field would be to design, with computer scientist Norman Margolus, a computer specifically to run cellular automata programs faster than even the supercomputers of the time. The Cellular Automata Machine (CAM) would help researchers ride the wave of interest triggered by Conway's Life. Complex patterns blossomed and faded in front of their eyes as the CAM whizzed through simulations, speeding discoveries they might have missed with their ponderous chuntering lab machines. In 1982, Fredkin organized an informal symposium to nurture the new field at his idyllic Caribbean retreat, Moskito Island (nowadays owned by British tycoon Richard Branson). But there was trouble in paradise, in the shape of a young mathematician by the name of Stephen Wolfram.

Wolfram cuts a divisive figure in the scientific world. He won a scholarship to Eton but never graduated. He went to Oxford University but, appalled by the standard of lectures, he dropped out. His next stop was the California Institute of Technology (CalTech), where he completed a PhD in theoretical physics. He was still only twenty. He joined the IAS in 1983 but left academia four years later after founding Wolfram Research. The company's flagship product, Mathematica, is a powerful tool for technical computing, written in a language he designed. Since its launch in 1988, millions of copies have been sold.

Wolfram began to work seriously on automata theory at the IAS. His achievements in the field are lauded – not least by himself – and he is dismissive of his predecessors. 'When I started,' he told journalist Steven Levy, 'there were maybe 200 papers written on cellular automata. It's amazing how little was concluded from those 200 papers. They're really bad.'[37]

Like Fredkin, Wolfram thinks that the complexity of the natural world arises from simple computational rules – possibly just one rule – executed repeatedly.[38] Wolfram guesses that a single cellular automaton cycling through this rule around 10^{400} times would be sufficient to reproduce all known laws of physics.[39] There is, however, one thing the two scientists disagree about: who came up with the ideas first.[40]

Fredkin insists he discussed his theories of digital cosmogenesis with Wolfram at the meeting in the Caribbean, and there catalysed his incipient interest in cellular automata. Wolfram says he discovered automata independently, and only afterwards found the work of von Neumann and others describing similar phenomena to those appearing on his computer screen. He traces his interest in automata back to 1972 – when he was twelve years old.[41] Sadly, Wolfram says, he did not take his ideas further then because he was distracted by his work on theoretical particle physics. Such assertions would be laughable had Wolfram not, in fact, begun publishing on quantum theory in respectable academic journals from 1975, aged fifteen.

Wolfram's first paper on cellular automata only appeared in 1983,[42] after the Moskito Island meeting, though he says he began work in the field two years earlier.[43] He was investigating complex patterns in nature, and was intrigued by two rather disparate questions: how do gasses coalesce to form galaxies, and how might McCulloch and Pitts-style model neurons be assembled into massive artificial neural networks? Wolfram started experimenting with simulations on his computer and was amazed to see strikingly ornate forms appear from computer code a few lines long. He forgot about galaxies and neural networks and started exploring the properties of the curious entities that he had stumbled upon.

The variety of automata Wolfram began playing with were one-dimensional. While von Neumann and Conway's constructions lived on two-dimensional surfaces, Wolfram's occupied a horizontal line. Each cell has just two neighbours, one on either side, and is either dead or alive. As with Conway's Game of Life, the easiest way to understand what Wolfram's automata are doing is by simulating their behaviour on a computer, with black and white squares for live and dead cells. Successive generations are shown one beneath the other, with the first, progenitor row at the top. As Wolfram's programs ran, a picture emerged on his screen that showed, line by line, the entire evolutionary history of his automata. He called them 'elementary cellular automata' because the rules that determined the fate of a cell were very straightforward.

In Wolfram's scheme, a cell communicates only with its immediate neighbours, and their states together with its own determine whether it will live or die. Those three cells can be in one of eight possible combinations of states. If '1' represents a live cell and '0' a dead one, these are: 111, 110, 101, 100, 011, 010, 001 and 000. For each of these configurations, a rule dictates what the central cell's state will be next. These eight rules, together with the starting line-up of cells, can be used to calculate every subsequent configuration of an elementary cellular automaton. Wolfram invented a numbering convention so that all eight rules governing an automaton's behaviour could be captured in a single number, from 0 to 255. Converting the decimal number to binary gives an eight-digit number that reveals whether the middle cell of each trio will be a '1' or a '0' next.

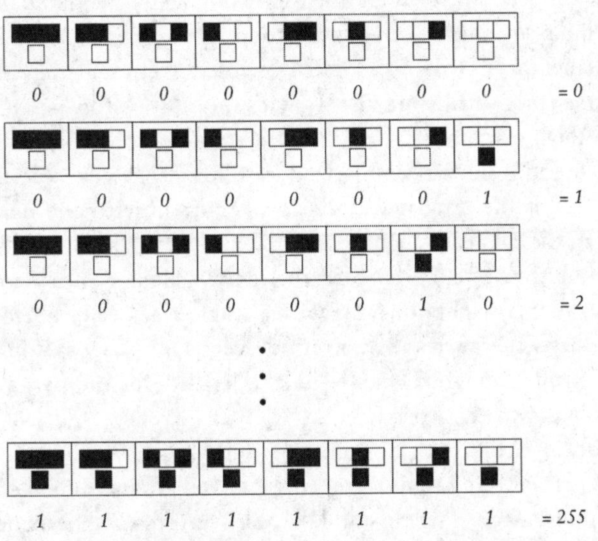

Wolfram's numbering convention for his elementary cellular automata.

Wolfram systematically studied all 256 different sets of rules available to his elementary cellular automata, sorting them carefully into groups like a mathematical zoologist.

Some of the rule sets produced very boring results. Rule 255 maps all eight configurations onto live cells, resulting in a dark screen; Rule 0 condemns all cells to death, producing a blank screen. Others resulted in repetitive patterns, for example Rule 250, which maps to the binary number 11111010. Starting from a single black cell, this leads to a widening checkerboard pattern.

A more intricate motif appears when an automaton follows Rule 90. The picture that emerges is a triangle composed of triangles, a sort of repetitive jigsaw puzzle, in which each identical piece is composed of smaller pieces, drawing the eye into its vertiginous depths. This kind of pattern, which looks the same at different scales, is known as a fractal, and this particular one, a Sierpiński triangle, crops up in a number of different cellular automata. In Life, for example, all that is required for the shape to appear is to start the game with a long line of live cells.

Still, pretty as these patterns are, they are hardly surprising. Looking at them, one could guess that a simple rule applied many times over might account for their apparent complexity. Would all elementary cellular automata turn out to produce patterns that were at heart also elementary? Rule 30 would show decisively that they did not. The same sorts of rule are at play with this automaton as with Wolfram's others – each describes the new colour of the central cell of a triad. The rules can be summarized as follows. The new colour of the central cell is the opposite of that of its current left-hand neighbour except in the two cases where both the middle cell and its right-hand neighbour are white. In this case, the centre cell takes the same colour as its left-hand neighbour. Start with a single black cell as before, and within fifty generations the pattern descends into chaos. While there are discernible diagonal bands on the left-side of the pyramid, the right side is a froth of triangles.

Wolfram checked the column of cells directly below the initial black square for any signs of repetition. Even after a million steps, he found none. According to standard statistical tests, this sequence of black and white cells is perfectly random. Soon after Wolfram presented his findings, people began sending him seashells with triangular markings that bore an uncanny resemblance to the 'random' pattern generated by Rule 30. For Wolfram, it was a strong hint that apparent randomness in nature, whether it appears on conch shells or in quantum physics, could be the product of basic algorithms just a few lines long.

Wolfram unearthed even more remarkable behaviour when he explored Rule 110. Starting from a single black square as before, the pattern expands only to the left, forming a growing right-angled triangle (rather than a pyramid like Rule 30) on which stripes shifted and clashed on a background of small white triangles. Even when Wolfram started with a random sequence of cells, ordered structures emerged and 'moved' left or right through the lattice. These moving

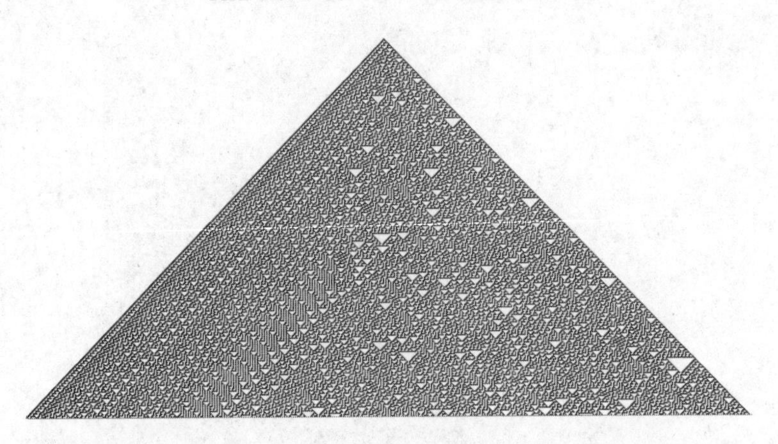

Rule 30.

The *Conus textile*, a venomous sea snail, has shell markings reminiscent of Rule 30.

patterns suggested an intriguing possibility. Were these regularities capable of carrying signals from one part of the pattern to another, like the 'gliders' in Conway's Life? If so, this incredibly rudimentary one-dimensional machine might also be a universal computer.

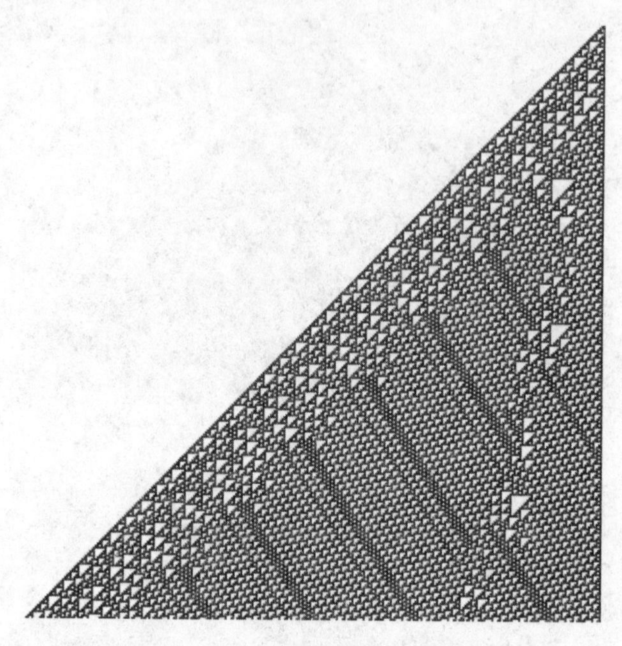

Rule 110

Mathematician Matthew Cook, who worked as Wolfram's research assistant in the 1990s, would prove it was.[44] Given enough space and time and the right inputs, Rule 110 will run any program – even *Super Mario Bros*.

Wolfram developed a classification system for automata based on computer experiments of these kinds. Class 1 housed automata which, like rule 0 or 255, rapidly converged to a uniform final state no matter what the initial input. Automata in Class 2 may end up in one of a number of states. The final patterns are all either stable or repeat themselves every few steps, like the fractal produced by Rule 90. In Class 3 are automata that, like Rule 30, show essentially random behaviour. Finally, Class 4 automata, like Rule 110, produce disordered patterns punctuated by regular structures that move and interact. This class is home to all cellular automata capable of supporting universal computation.[45]

The stream of papers that Wolfram was producing on automata was causing a stir, but not everyone approved. Attitudes at the IAS

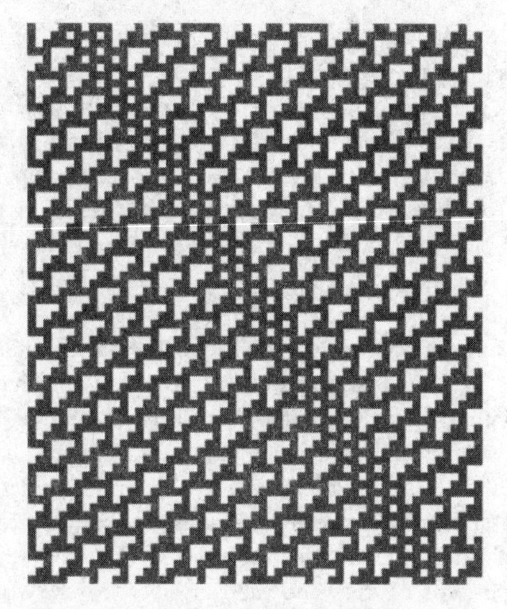

Detail from Rule 110 pattern showing a 'glider'.

had not moved on much from von Neumann's day, and the idea that meaningful mathematics could be done on computers – Wolfram's office was chock full of them – was still an anathema. Wolfram's four-year post came to an end in 1986, and he was not offered the chance to join the illustrious ranks of the permanent staff. A year later, busy with his new company he paused his foray into the world of cellular automata – only to stage a dramatic return to the field in 2002, with *A New Kind of Science*.[46] That book, the result of ten years of work done in hermit-like isolation from the wider scientific community, is Wolfram's first draft of a theory of everything.

The 1,280-page tome starts with characteristic modesty. 'Three centuries ago science was transformed by the dramatic new idea that rules based on mathematical equations could be used to describe the natural world,' he declares. 'My purpose in this book is to initiate another such transformation.' And that transformation, he explained, would be achieved by finding the single 'ultimate rule' underlying all other physical laws – the automaton to rule them all, God's four-line computer program. Wolfram had not found that rule. He was not even close. But he did tell a journalist he expected the code to be found in his lifetime – perhaps by himself.[47]

The book presents in dazzling detail the results of Wolfram's painstaking investigations of cellular automata of many kinds. His conclusion is that simple rules can yield complex outputs but adding more rules (or dimensions) rarely adds much complexity to the final outcome. In fact, the behaviour of all the systems he surveys falls into one of the four classes that he had identified years earlier. Wolfram then contends similar simple programs are at work in nature and marshals examples from biology, physics and even economics to make his case. He shows, for example, how snowflake-like shapes can be produced with an automaton that turns cells on a hexagonal lattice black only when exactly one neighbour is black. Later, he simulates the whorls and eddies in fluids using just five rules to determine the outcome of collisions between molecules moving about on a two-dimensional grid. Similar efforts produce patterns resembling the branching patterns of trees and the shapes of leaves. Wolfram even suggests the random fluctuations of stocks on short timescales might be a result of simple rules and shows a one-dimensional cellular automaton roughly reproduces the sort of spiking prices seen in real markets.

The most ambitious chapter contains Wolfram's ideas on fundamental physics, encapsulating his thoughts on gravity, particle physics and quantum theory. He argues that a cellular automaton is not the right choice for modelling the universe. Space would have to be partitioned into tiny cells, and some sort of a timekeeper would be required to ensure that every cell updates in perfect lockstep. Wolfram's preferred picture of space is a vast, sub-microscopic network of nodes, in which each node is connected to three others. The sort of automaton that he thinks underlies our complex physics is one that changes the connections between neighbouring nodes according to some simple rules – for instance, transforming a connection between two nodes into a fork between three. Apply the right rule throughout the network billions of times and, Wolfram hopes, the network will warp and ripple in exactly the ways predicted by Einstein in his general theory of relativity. At the opposite end of the scale, Wolfram conjectures that elementary particles like electrons are the manifestations of persistent patterns rippling through the network – not so different from the gliders of rule 110 or Life.

The problem with Wolfram's models, as his critics were quick to point out, is that it is impossible to determine whether they reflect reality unless they make falsifiable predictions – and Wolfram's do not (yet). Wolfram hoped that his book would galvanize researchers to adopt his methods and push them further, but his extravagant claim of founding 'a new kind of science' got short shrift from many scientists. 'There's a tradition of scientists approaching senility to come up with grand, improbable theories,' Freeman Dyson told *Newsweek* after the book's release. 'Wolfram is unusual in that he's doing this in his 40s.'[48]

Disheartened by the reception of the book in scientific circles, Wolfram disappeared from the public eye only to re-emerge almost twenty years later, in April 2020.[49] With the encouragement of two young acolytes, physicists Jonathan Gorard and Max Piskunov, his team at Wolfram Research had explored around a thousand universes that emerged from following different rules.[50] Wolfram and his colleagues showed that the properties of these universes were consistent with some features of modern physics.

The results intrigued some physicists, but the establishment, in general, was once again unenthusiastic. Wolfram's habit of belittling or obscuring the accomplishments of his predecessors has garnered a less receptive audience than his ideas otherwise might have had. 'John von Neumann, he absolutely didn't see this,' Wolfram says, insisting that he was the first to truly understand that enormous complexity can spring from automata. 'John Conway, same thing.'[51]

A New Kind of Science is a beautiful book. It may prove in time also to be an important one. The jury, however, is still very much out. Whatever the ultimate verdict of his peers, Wolfram had succeeded in putting cellular automata on the map like no one else. His groundwork would help spur those who saw automata not just as crude simulations of life, but as the primitive essence of life itself.

'If people do not believe that mathematics is simple,' von Neumann once said, 'it is only because they do not realize how complicated life is.'[52] He was fascinated by evolution through natural selection, and some of the earliest experiments on his five-kilobyte IAS machine involved strings of code that could reproduce and mutate like DNA. The man who sowed those early seeds of digital life was an eccentric, uncompromising Norwegian-Italian mathematician named Nils Aall Barricelli.[53]

A true maverick, Barricelli was never awarded a doctorate because he refused to cut his 500-page long thesis to a length acceptable to his examiners. He was often on the wrong side of the line that separates visionaries from cranks, once paying assistants out of his own not-very-deep pockets to find flaws in Gödel's famous proof. He planned, though never finished, a machine to prove or disprove mathematical theorems. Some of his ideas in biology, however, were genuinely ahead of their time. Von Neumann had a proclivity for helping people with big ideas. When Barricelli contacted him to ask for time on the IAS machine, he was intrigued. 'Mr. Barricelli's work on genetics, which struck me as highly original and interesting . . . will require a great deal of numerical work,' von Neumann wrote in support of Barricelli's grant application, 'which could be most advantageously effected with high-speed digital computers of a very advanced type.'

Barricelli arrived in Princeton in January 1953, loosed his numerical organisms into their digital habitat on the night of 3 March 1953 and birthed the field of artificial life. Barricelli believed that mutation and natural selection alone could not account for the appearance of new species. A more promising avenue for that, he thought, was symbiogenesis, in which two different organisms work so closely together that they effectively merge into a single, more complex creature. The idea of symbiogenesis had been around since at least the turn of the twentieth century. For passionate advocates like Barricelli, the theory implies that it is principally cooperation not competition between species that drives evolution.

The micro-universe Barricelli created inside von Neumann's machine was designed to test his symbiogenesis-inspired hypothesis that genes themselves were originally virus-like organisms that had, in the past, combined and paved the way to complex, multi-cellular organisms. Each 'gene' implanted in the computer's memory was a number from negative to positive 18, assigned randomly by Barricelli using a pack of playing cards.[54] His starting configuration was a horizontal row of 512 genes. In the next generation, on the row below, each gene with number, *n*, produces a copy of itself *n* places to the right if it is positive, *n* places to the left if it is negative. The gene '2', for example, occupies the same position on the row beneath it, but also produces a copy of itself two places to the right. This was the act of reproduction. Barricelli then introduced various rules to simulate mutations. If two numbers try to occupy the same square, for example, they are added together and so on. 'In this manner,' he wrote, 'we have created a class of numbers which are able to reproduce and to undergo hereditary changes. The conditions for an evolution process according to the principle of Darwin's theory would appear to be present.'[55]

To simulate symbiogenesis, Barricelli changed the rules so that the 'genes' could *only* reproduce themselves with the help of another of a different type. Otherwise, they simply wandered through the array, moving left or right but never multiplying.

The IAS machine was busy with bomb calculations and weather forecasts, so Barricelli ran his code at night. Working alone, he cycled punched cards through the computer to follow the fate of his species over thousands of generations of reproduction. 'No process of evolution had ever been observed prior to the Princeton experiments,' he proudly told von Neumann.[56]

But the experiments performed in 1953 were not entirely successful. Simple configurations of numbers would invade to become the single dominant organism. Other primitive combinations of numbers acted like parasites, gobbling up their hosts before dying out themselves. Barricelli returned to Princeton a year later to try again. He tweaked the rules, so that, for instance, different mutation rules pertained in different sections of each numerical array. Complex coalitions of numbers propagated through the matrix, producing patterns resembling the outputs of Wolfram's computer simulations decades later. That is no coincidence – Barricelli's numerical organisms were effectively one-dimensional automata too. On this trip, the punched cards documented more interesting phenomena, which Barricelli likened to biological events such as self-repair and the crossing of parental gene sequences.

Barricelli concluded that symbiogenesis was all important and

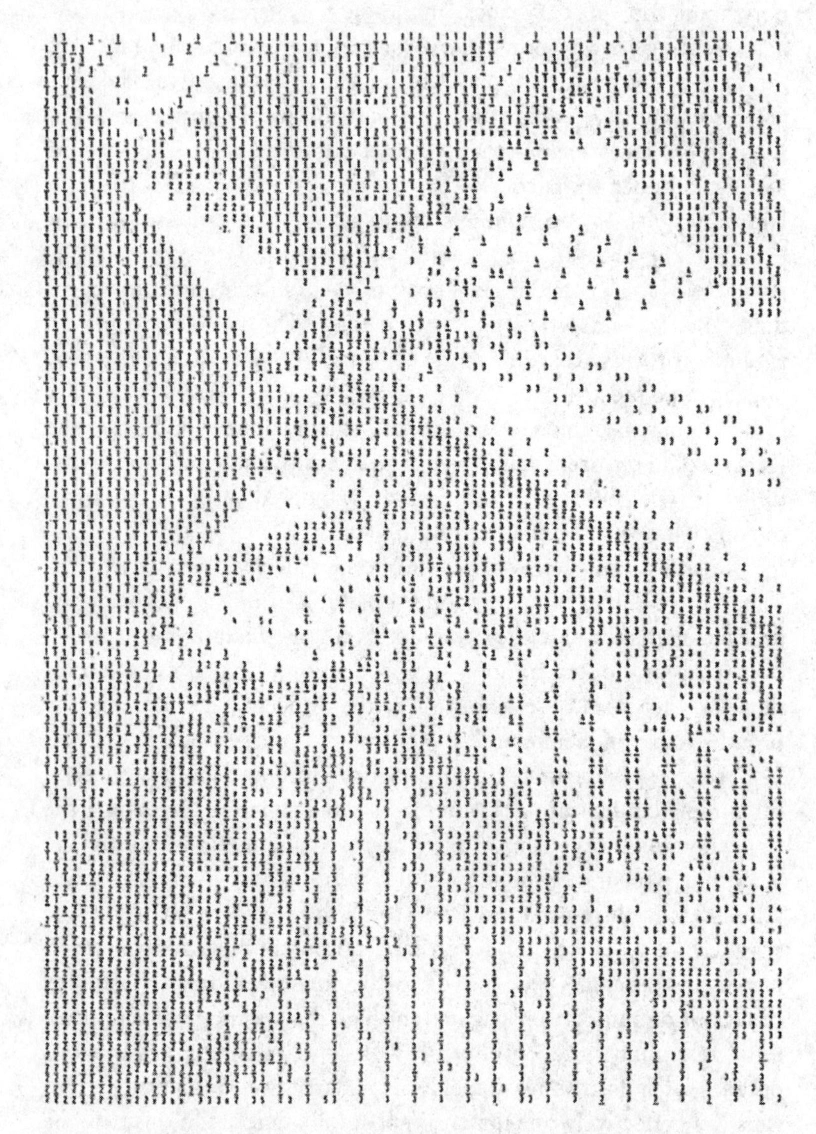

'Spontaneous formation of symbioorganisms in an experiment started with random numbers'. *From Nils Aall Barricelli, 'Numerical Testing of Evolution Theories. Part I: Theoretical Introduction and Basic Tests'*, Acta Biotheoretica, *16 (1962), pp. 69–98.*

omnipresent, perhaps on extrasolar planets as well as on Earth. Random genetic changes coupled with natural selection, he argued, cannot account for the emergence of multi-cellular life forms. On that he was wrong, but symbiogenesis is now the leading explanation of how plant and animal cells arose from simpler, prokaryotic organisms. The recent discovery of gene transfer between a plant and an animal (in this case the whitefly *Bemisia tabaci*) suggests this process is more widespread than was recognized in Barricelli's day.[57]

In the sixties, Barricelli evolved numerical organisms to play 'Tac-Tix', a simple two-player strategy game played on a four-by-four grid, then, later, chess. The algorithms he developed anticipated machine learning, a branch of artificial intelligence that focuses on applications that improve at a task by finding patterns in data. But perhaps because he was so out of step with mainstream thought, Barricelli's pioneering work was largely forgotten. Only decades later would the study of numerical organisms be properly revived. The first conference dedicated to artificial life would be held at Los Alamos in September 1987, uniting under one banner scientists working in disciplines as disparate as physics, anthropology and genetics (evolutionary biologist Richard Dawkins was one of around 160 attendees). The 'Interdisciplinary Workshop on the Synthesis and Simulation of Living Systems' was convened by computer scientist Christopher Langton, who expressed the aims of the field in terms that would have been familiar to Barricelli. 'Artificial life is the study of artificial systems that exhibit behaviour characteristic of natural living systems ... The ultimate goal,' Langton wrote, 'is to extract the logical form of living systems.'[58] Los Alamos had birthed in secret the technology of death. Langton hoped that arid spot would also be remembered one day for birthing new kinds of life.

Langton first encountered automata while programming computers in the early 1970s for the Stanley Cobb Laboratory for Psychiatric Research in Boston. When people started running Conway's Life on the machines, Langton was enthralled and began experimenting with the game himself. His thoughts on artificial life, however, did not begin to coalesce until he was hospitalized in 1975 following a near-fatal hang-gliding accident in North Carolina's Blue Ridge Mountains. With plenty of time to read, Langton, who had no formal scientific training, recognized where his true interests lay. The following year, he enrolled at the University of Arizona in Tucson and assembled from the courses on offer his own curriculum on the principles of artificial biology. When the first personal computers appeared, he bought one and began trying to simulate evolutionary processes. It was then, while trawling through the library's shelves, that Langton discovered von Neumann's work on self-reproducing automata and hit upon the idea of simulating one on his computer.

Langton quickly realized that the 29-state cellular automaton was far too complicated to reproduce with the technology he had to hand. He wrote to Burks, who informed him that Edgar Codd, an English computer scientist, had simplified von Neumann's design a few years earlier, as part of doctoral work at the University of Michigan. Codd's 8-state automaton was, like von Neumann's, capable of universal computation and construction. When Langton found that Codd's machine was also far too complex to simulate on his Apple II, he dropped those requirements. 'It is highly unlikely,' he reasoned later, 'that the earliest self-reproducing molecules, from which all living organisms are supposed to have been derived, were capable of universal construction, and we would not want to eliminate those from the class of truly self-reproducing configurations.'[59]

What Langton arrived at after many rounds of simplification were what he called 'loops'. Each loop is a square-ish doughnut with a short arm jutting out from one corner, resembling a prostrate letter 'p'. When a loop begins its reproductive cycle, this appendage grows, curling around to form an identical square-shaped loop next to its parent. On completion, the link to the parent is severed, and parent and daughter each grow another arm from a different corner to begin another round of replication. Loops surrounded on all four sides by other loops are unable to germinate new arms. These perish, leaving a dead core surrounded by a fertile outer layer, like a coral reef.

Determined to pursue his thoughts on automata further, Langton searched for a graduate programme that would accept him. He found only one: the Logic of Computers Group founded by Burks. In 1982, aged thirty-three, unusually mature for a graduate student, Langton arrived at the University of Michigan ready to begin his studies. He soon encountered Wolfram's early research on classifying automata

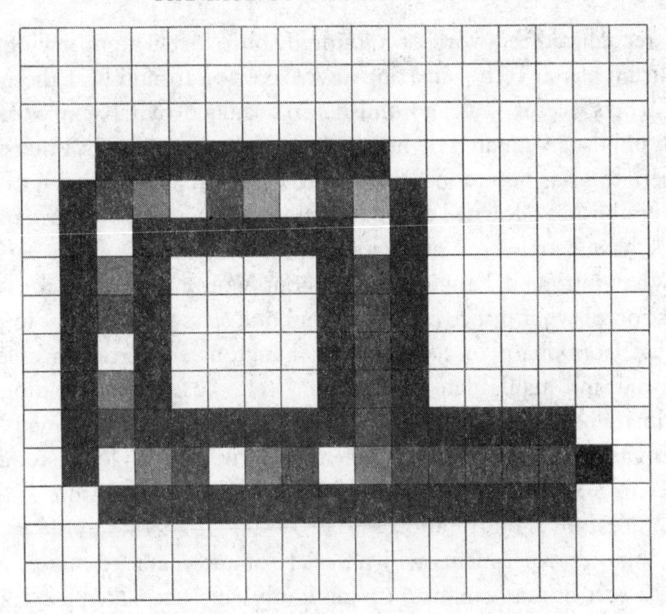

A Langton loop.

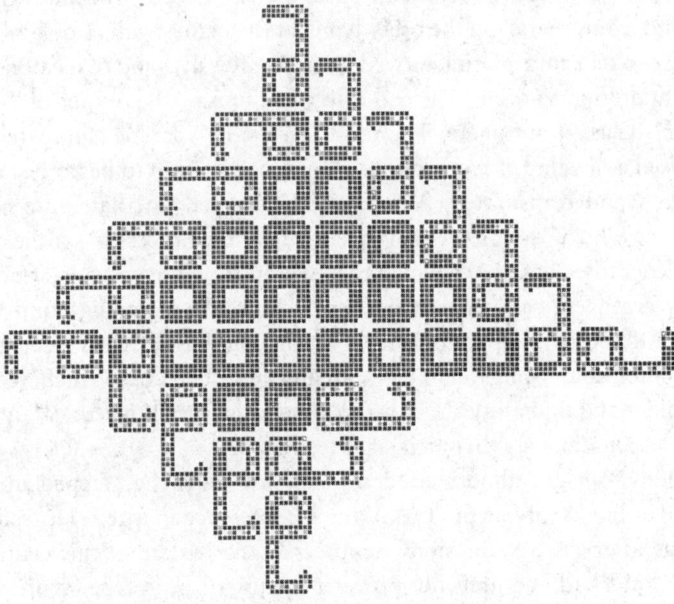

A colony of loops.

and recognized the work of a kindred spirit. 'After a paper by Steve Wolfram of Cal Tech – amazing how I screwed around with these linear arrays over a year ago and never imagined that I was working with publishable material,' he wrote in his journal. 'Surely, I thinks to myself, this has been done 30 years ago, but No! Some kid out of Cal Tech publishes a learned paper on one dimensional two-state arrays!! What have people been *doing* all these years?'[60]

What intrigued Langton most about Wolfram's system was the question of what tipped a system from one class to another – in particular, from chaos to computation. Langton felt automata able to transmit and manipulate information were closest to the biological automata he wanted to emulate. 'Living organisms,' he explained, 'use information in order to rebuild themselves, in order to locate food, in order to maintain themselves by retaining internal structure . . . the structure itself is information.'[61]

Langton's experiments with a variety of automata led him to propose a sort of mathematical tuning knob which he called a system's 'lambda (λ) parameter'.[62] On settings of lambda close to zero, information was frozen in repetitive patterns. These were Wolfram's Class 1 and 2 automata. On the other hand, at the highest values of lambda, close to one, information moved *too* freely for any kind of meaningful computation to occur. The end result was noise – the output of Wolfram's Class 3 automata. The most interesting Class 4 automata are poised at a value of lambda that allows information to be both stably stored and transmitted. All of life is balanced on that knife edge 'below which,' as von Neumann had put it, 'the process of synthesis is degenerative, but above which the phenomenon of synthesis, if properly arranged, can become explosive'. And if that tipping point had been found by happy accident on Earth, could it be discovered again by intelligent beings actively searching for it? Langton believed it could – and humanity should start preparing for all the ethical quandaries and dangers that artificial life entailed.

Langton had almost cried tears of joy when he stepped up to deliver his lecture at the Los Alamos conference in 1987. His reflections afterwards were more sombre. 'By the middle of this century, mankind had acquired the power to extinguish life,' he wrote. 'By the end of the century, he will be able to create it. Of the two, it is hard to say which places the larger burden of responsibilities on our shoulders.'[63]

Langton's prediction was a little off. Mankind had to wait a decade longer than he envisaged for the first lab-made organism to appear. In 2010, the American biotechnologist and entrepreneur Craig Venter and his collaborators made a synthetic near-identical copy of a genome from the bacterium *Mycoplasma mycoides* and transplanted it into a cell that had its own genome removed.[64] The cell 'booted up' with the new instructions and began replicating like a natural bacterium. Venter hailed 'the first self-replicating species we've had on the planet whose parent is a computer' and though plenty of scientists disagreed that his team's creation was truly a 'new life form', the organism was quickly dubbed 'Synthia'.[65] Six years later, researchers from Venter's institute produced cells containing a genome smaller than that of any independently replicating organism found in nature. They had pared down the *M. mycoides* genome to half its original size through a process of trial and error. With just 473 genes, the new microorganism is a genuinely novel species, which they named 'JCVI-syn3.0a'.[66] The team even produced a computer simulation of their beast's life processes.[67]

Others hope to go a step further. Within a decade, these scientists hope to build synthetic cells from the bottom up by injecting into oil bubbles called liposomes the biological machinery necessary for the cells to grow and divide.[68]

Von Neumann's cellular automata seeded grand theories of everything and inspired pioneers who dared to imagine life could be made from scratch. His unfinished kinematic automaton too has borne fruit. Soon after Kemeny publicized von Neumann's ideas, other scientists wondered if something like it could be realized not in computer simulations, but in real-life physical stuff. Rather than the 'soft' world of biology, their devices were rooted in the 'hard' world of mechanical parts, held together with nuts and bolts – 'clanking replicators' as nanotechnology pioneer Eric Drexler dubbed them, or the autofacs of Dick's story.[69]

Von Neumann had left no blueprint for anyone to follow – his automaton was a theoretical construct, no more than a thought experiment to prove that a machine could make other machines including itself. Consequently, the earliest efforts of scientists to realize kinetic automata were crude, aiming only to show that the ability to reproduce was not exclusively a trait of living things. As geneticist Lionel Penrose noted, the very idea of self-reproduction was so 'closely associated with the fundamental processes of biology that it carries with it a suggestion of magic'.[70] With his son, future Nobel laureate Roger, Penrose made a series of plywood models to show that replication could occur without any hocus-pocus. Each set of wooden blocks had an ingenious arrangement of catches, hooks and levers that were designed to allow only certain configurations to stick together. By shaking the blocks together on a tray, Penrose was able to demonstrate couplings and cleavages that he likened to natural processes such as cell division. One block, for instance, could 'catch' up to three others. This chain of four would then split down the middle to form a pair of new two-block 'organisms', each of which could grow and reproduce like their progenitor.

Chemist Homer Jacobson performed a similar sort of trick using a modified model train set.[71] His motorized carriages trundled around a circular track, turning into sidings one by one to couple with others looking for all the world as if they knew exactly what they were doing. With only the aid of a children's toy, Jacobson had conjured up a vision of self-replicating machines.

Clanking replicators.

Then there were dreamers who were not content with merely demonstrating the principles of machine reproduction. They wanted to tap the endless production potential of clanking replicators. What they envisioned, in essence, were benign versions of Dick's autofac. One of the first proposals sprang from the imagination of mathematician Edward F. Moore. His 'Artificial Living Plants' were solar-powered coastal factories that would extract raw materials from the air, land and sea.[72] Following the logic of von Neumann's kinematic automaton, the plants would be able to duplicate themselves any number of times. But their near-limitless economic value would come from whatever other products they would be designed to manufacture. Moore thought that fresh water would be a good 'crop' in the first instance. He envisaged the plants migrating periodically 'like lemmings' to some pre-programmed spot where they would be harvested of whatever goods they had made.

'If the model designed for the seashore proved a success, the next step would be to tackle the harder problems of designing artificial living plants for the ocean surface, for desert regions or for any other locality having much sunlight but not now under cultivation,' said Moore. 'Even the unused continent of Antarctica,' he enthused, 'might be brought into production.'

Perhaps unnerved by the prospect of turning beaches and pristine tracts of wilderness into smoggy industrial zones, no one stepped forward to fund Moore's scheme. Still, the idea of getting something for nothing (at least, once the considerable start-up costs are met) appealed to many, and the concept lingered on the fringes of science. Soon, minds receptive to the idea of self-reproducing automata thought of a way to overcome one of the principal objections to Moore's project. The risk that the Earth might be swamped by machines intent on multiplying themselves endlessly could be entirely mitigated, they conjectured – by sending them into space. In recognition of the fact these types of craft are inspired by the original theory of self-reproducing automata, they are now called von Neumann probes.

Among the first scientists to think about such spacecraft was von Neumann's IAS colleague Freeman Dyson. In the 1970s, building on Moore's idea, he imagined sending an automaton to Enceladus, the sixth-largest of Saturn's moons.[73] Having landed on the moon's icy surface, the automaton churns out 'miniature solar sailboats', using only what materials it can garner. One by one, each sailboat is loaded with a chunk of ice and catapulted into space. Buoyed by the pressure of sunlight on their sails, the craft travel slowly to Mars, where they deposit their cargo, eventually depositing enough water to warm the Martian climate and bring rain to the arid planet for the first time in a billion years. A less fantastic proposal was put forward by Robert Freitas, a physicist who also held a degree in law. In 2004, he and computer scientist Ralph Merkle would produce the veritable bible of self-replicating technology, *Kinematic Self-Replicating Machines*, a definitive survey of all such devices, real or imagined.[74] In the summer of 1980, Freitas made a bold contribution to the field himself with a probe designed to land on one of Jupiter's moons and produce an interstellar starship, dubbed 'REPRO', once every 500 years.[75] This might seem awkwardly long to impatient Earth-bound mortals, but since REPRO's ultimate purpose is to explore the galaxy, there was no particular hurry. Freitas estimated that task would take around 10 million years.

The most ambitious and detailed proposal for a space-based replicator was cooked up over ten weeks in a Californian city located in the heart of Silicon Valley. In 1980, at the request of President Jimmy Carter, NASA convened a workshop in Santa Clara on the role of artificial intelligence and automation in future space missions. Eighteen academics were invited to work with NASA staff. By the time the final report was filed, the exercise had cost over US$11 million.

The group quickly settled on four areas that they thought would require cutting-edge computing and robotics, then split into teams to flesh out the technical requirements and goals of each mission. The ideas included an intelligent system of Earth observation satellites, autonomous spacecraft to explore planets outside the Solar System and automated space factories that would mine and refine materials from the Moon and asteroids. The fourth proposal was considered the most far-fetched of all. Led by Richard Laing, the team behind it laid out how a von Neumann-style automaton might colonize the Moon, extraterrestrial planets and, in time, the far reaches of outer space. 'Replicating factories should be able to achieve a very general manufacturing capability including such products as space probes, planetary landers, and transportable 'seed' factories for siting on the surfaces of other worlds,' they declared in their report. 'A major benefit of replicating systems is that they will permit extensive exploration and utilization of space without straining Earth's resources.'[76]

Laing was another alumnus of Burks's group. He had arrived at the University of Michigan after dropping out of an English literature degree to work as a technical writer for some of the computer scientists based there. When Burks founded the Logic of Computers Group, Laing was drawn to the heady debates on machine reproduction. He soon joined the group and, after finishing a doctorate, explored the implications of von Neumann's automata. One of Laing's most lasting contributions would be his proof that an automaton need not start with a complete description of itself in order to replicate. He showed that a machine that was equipped with the means to inspect itself could produce its own self-description and so reproduce.[77]

At the NASA-convened workshop in Santa Clara, Laing soon found kindred spirits. As well as Freitas, who had recently outlined his own self-reproducing probe design, Laing was joined by NASA engineer Rodger Cliff and German rocket scientist Georg von Tiesenhausen. Tiesenhausen had worked with Wernher von Braun on the V-2 programme during the Second World War. Brought to the United States, he helped design the Lunar Roving Vehicle for the Apollo moon missions.

The team's ideas were controversial, and they knew it. To dispel the whiff of crankery surrounding their project, they marshalled every bit of science in their favour. While other teams began their reports by stressing the benefits of their chosen missions, the 'Self-Replicating Systems' (SRS) group started defensively, carefully laying out the theoretical case that what they proposed was possible at all. 'John von Neumann', they conclude sniffily, 'and a large number of other researchers in theoretical computer science following him have shown that there are numerous alternative strategies by which a machine system can duplicate itself.'

The SRS team produced two detailed designs for fully self-replicating lunar factories. The first unit is a sprawling manufacturing hub that strip-mines surrounding land to make commercial products or new copies of itself. A central command and control system orchestrates the whole operation. Extracted materials are analysed and processed into industrial feedstock and stored in a materials depot. A parts production plant uses this feedstock to make any and all components the factory needs. These parts are then either transported to a facility to make whatever products the Earth commands or to a universal constructor, which assembles more factories.

A drawback of this scheme is that a whole factory has to be built on the Moon before the automaton can replicate. The team's second design, a 'Growing Lunar Manufacturing Facility' avoids this difficulty by requiring nothing more to start construction than a single 100 ton spherical 'seed' craft, packed with robots dedicated to different tasks. Dropped onto the lunar surface, the seed cracks open to release its cargo. Once more, a master computer directs the action. First scouting bots survey the immediate surroundings of the seed to establish where exactly the facility should be built. A provisional solar array is erected to provide power. Five paving robots roll out of the craft and construct solar furnaces to melt lunar soil, casting the molten rock into basalt slabs. The slabs are laid down to form the factory's foundations, a circular platform 120 metres in diameter. Working in parallel, other robots begin work on a massive roof of solar cells that will eventually cover the entire workspace, supplying the unit with all the power it needs for manufacturing and self-replication. Meanwhile, sectors for chemical processing, fabrication and assembly are set up. Within a year of touchdown, the team predicted, the first self-replicating factory on the Moon will be fully functional, churning out goods and more factories.

Like Alex Ellery many years later, Laing's team worried about 'closure' – could their automaton operate with the materials and energy available? They concluded that the Moon would supply about 90 per cent of the factory's needs. Unlike Ellery, who is shooting for 100 per cent closure with his hybrid of 3D printer and rover, they accepted that at the outset of the project, the remaining 4–10 per cent would have to be sent from Earth. These 'vitamin parts', they said, 'might include hard-to-manufacture but lightweight items such as microelectronics components, ball bearings, precision instruments and others which may not be cost-effective to produce via automation off-Earth except in the longer term'.

The social and philosophical problems of their project did not escape the team either. They expected their factories to evolve, just like a biological replicator. Left to their own devices in space, would these machines become conscious, and if so, would they necessarily be happy to serve our ends rather than their own? The machines could be taught 'right' and 'wrong', but as with people, there was no guarantee of good behaviour. This unsettling conclusion led them to ask if humans would always be able switch off the autonomous factories should they pose a threat – the so-called 'unpluggability problem'. They thought not. 'At some point the depth of analysis and sophistication of action available to a robot system may exceed the abilities of human minds to defeat it,' they say. How to prevent such machines from becoming 'unpluggable' should be an 'urgent subject for further research'.

Yet what Laing's team felt most keenly was the boundless potential of von Neumann's automata. 'How will humankind deal with what has been termed, with some justification, "the last machine we need ever build"?' they asked. Unleashing replicating machines, they said, may have 'implications on a cosmological scale'.

> Humanity could set in motion a chain reaction of organization sweeping across the Universe at near the speed of light. This organized part of the Universe could itself be viewed as a higher level 'organism.' Instead of merely following the laws of mechanics and thermodynamics, something unique in our knowledge would occur. The degree of cosmic organization would increase. Life could become commonplace, whereas now it seems quite rare. New rules, the rules of life, would spread far and wide.

It was not to be. When, in 1983, Laing heard rumours that President Ronald Reagan was about to launch a massive new space initiative, he looked forward to the speech with great anticipation. What the president announced instead was the Strategic Defense Initiative – 'Star Wars'.

The dream has not been forgotten. In 2021, the Initiative for Interstellar Studies, a charity based in London, unveiled an updated design of a von Neumann probe that they say could be built within a

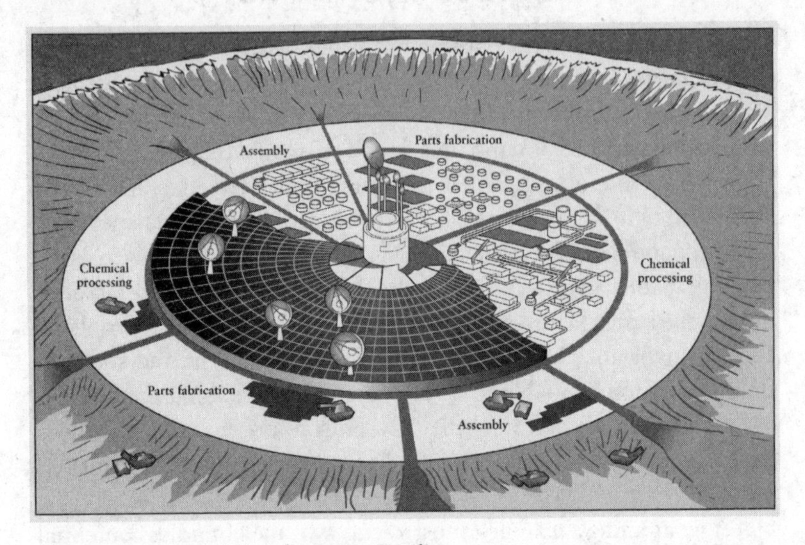

A Growing Lunar Manufacturing Facility.

decade.[78] Ellery's group is whittling away at the last few percentage points that will get them to 100 per cent closure. The faithful wait for the call from a patron truly prepared to 'dare mighty things'.

By the 1980s, von Neumann's name was synonymous with the idea of self-replicating machines. The headiest days of human space exploration were long gone, but a new era of genetic engineering had arrived, and molecular biology was in the ascendant. In 1982, the US Food and Drug Administration approved Genentech's 'Humulin', a version of the hormone insulin produced by bacteria. The first genetically modified plant, an antibiotic-resistant strain of tobacco, was reported the following year.[79] Scientists intrigued by the concept of replicators turned their thoughts from the cosmos to a submicroscopic world populated by molecular machines, rather than robots.

The first to articulate this vision was American engineer Eric Drexler, who began using the term 'nanotechnology' to describe the new field of molecular manufacturing. 'When biochemists need complex molecular machines, they still have to borrow them from cells,' Drexler says in his influential book *Engines of Creation*. 'Nevertheless, advanced molecular machines will eventually let them build nanocircuits and nanomachines as easily and directly as engineers now build microcircuits or washing machines.'[80]

The first nanomachines, Drexler predicts, would be built from proteins, which already perform mechanical tasks in cells. Happily, machines built from these biomolecules would not even need to be assembled. Much like the wooden replicators Penrose designed link up in precise configurations when shaken together, the protein parts of a complex structure like the outer shell of a virus bounce off each other until complementary chemical forces pull them together and make them stick. These soft biological machines will then serve as a stepping stone to second-generation nanomachines made of tougher stuff, such as ceramics, metals and diamond.

Drexler had come across the transcript of a provocative talk entitled 'There's Plenty of Room at the Bottom' given by Richard Feynman in 1959. 'What would happen', Feynman asks, 'if we could arrange the atoms one by one the way we want them?' Fired up by the idea, which was consistent with his own thinking at the time, Drexler wanted to realize Feynman's vision. In *Engines of Creation*, he proposes an 'assembler' a few billionths of a metre long that is able to both replicate itself and make other machines, atom by atom. And, to his later regret, [81] he warned of 'gray goo' – an apocalyptic scenario that follows the accidental release of dangerous replicators from a lab; 'they could spread like blowing pollen, replicate swiftly, and reduce the biosphere to dust in a matter of days'.[82]

Some eminent scientists dismissed the whole idea and accused Drexler of scaremongering. 'There will be no such monster as the self-replicating mechanical nanobot of your dreams,' Nobel-prize winning chemist Richard Smalley growled at Drexler from the pages of *Chemical & Engineering News*.[83]

But in September 2017, researchers at the University of Manchester published details in *Nature* of a molecule that could be programmed to make one of four different molecules.[84] More recently, chemists at the University of Oxford even claim to have made a rudimentary self-replicating assembler.[85] Though Drexler now downplays the role of self-replicating nanobots in molecular manufacturing, his speculations look less far-fetched than they once did.

*

Von Neumann's theory continued to spread, replicate and evolve much like the self-reproducing machines it describes. 'Agent-based models', which try to simulate the interaction of autonomous individuals, grew out of von Neumann's cellular automaton. Economist Thomas Schelling, no stranger to von Neumann's work, was one of the first to use the technique. He investigated segregation in cities with nothing more than a chessboard and two sets of coloured counters (let's say they are either pink or brown) representing people from two distinct groups.[86] The resulting 'game' resembles a two-dimensional automaton like Conway's Life.

His most famous experiment begins with equal numbers of pink and brown counters distributed at random across the board. About a quarter of the squares are left empty to allow counters to move around the board. Each set of counters is assumed to have the same fixed preference for neighbours of its own colour. This preference is expressed as a ratio of a counter's own colour to counters of the opposite colour in surrounding squares. Imagine that no counter is 'happy' if fewer than half its neighbours are of the same colour. Unhappy counters move to the nearest empty square that satisfies their preferences. Most squares have eight neighbours, so a pink counter with three or fewer pinks in surrounding squares will move. Pinks in edge squares move if fewer than three out of five neighbours are of like colour. For corner squares, a pink counter is not content unless at least two of three neighbours are pink. Likewise for brown counters.

When Schelling applied these rules, an initially mixed board partitioned quickly into homogeneous neighbourhoods of the same colour.[87]

Even a relatively mild proclivity for like-coloured neighbours resulted in segregation. Any preference stronger than 1:3 triggered exodus and eventual partition. 'Surprisingly, the results generated by this analysis do not depend upon each color's having a preference for living separately,' noted Schelling. 'They do not even depend on a preference for being in the majority!'[88]

Furthermore, once the counters had segregated, the process could not be reversed by softening the bias towards like-coloured companions. A return to more mixed neighbourhoods occurs only when an

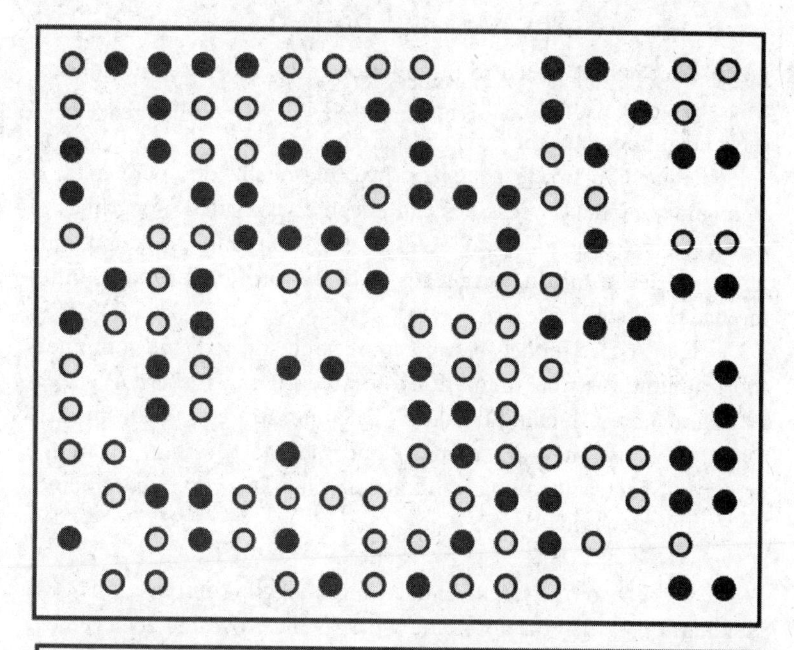

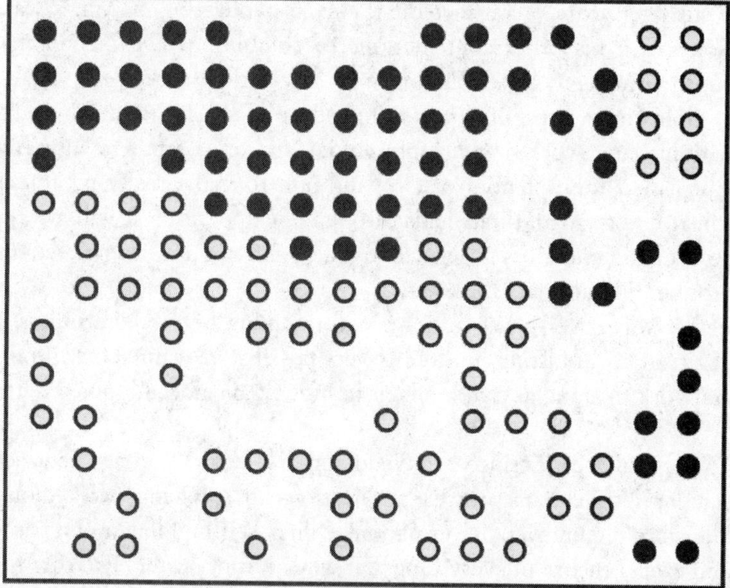

Starting configuration (top); end configuration (bottom). *From Thomas C. Schelling, 'Dynamic Models of Segregation',* Journal of Mathematical Sociology, *1 (1971), pp. 143–86.*

upper limit is introduced to the fraction of like-coloured neighbours that a counter wants. That is when, for example, counters move if all their neighbours are the same colour.

Schelling had produced two powerful conclusions with a fairly elementary model. First, cities can become segregated along lines of race even if no one minds living in a mixed community. Second, only an active desire for diversity leads to diverse neighbourhoods. Indifference results in segregation.

As agent-based models became more sophisticated, and computers more ubiquitous, the simulations became digital. Robert Axelrod's tournaments for duelling Prisoner's Dilemma strategies were an early example. Now, they are used to understand everything from the growth of bacterial colonies and housing markets to tax compliance and voting behaviour.

Perhaps because of its near endless mindboggling implications, von Neumann regarded his automata work as the crowning achievement of his later years. Ulam says the theory of replicating machines was, along with his other contributions to computing, von Neumann's 'most permanent, most valuable, most interesting work'.[89]

Goldstine agrees. 'It not only linked up his early interest in logic with his later work on neurophysiology and on computers,' he says, 'but it had the potential of allowing him to make really profound contributions to all three fields through one apparatus. It will always be a fundamental loss to science that he could not have completed his program in automata theory . . .'

'He was never given to bragging or staking out a claim unless it deserved it,' Goldstine continues. 'Very possibly he wanted his automata work to stand as a monument to himself, as indeed it does.'[90]

Over seven decades since von Neumann first lectured on his theory of cellular automata, its possible implications are still being worked out. Plausibly, it could yet give us nanomachines, self-building moon bases and even a theory of everything. Yet while it took just a few years for Turing's computing machine to be turned from mathematical abstraction to physical reality, the self-replicating machines von Neumann imagined have not yet been made. Or have they?

'The computer – a new form of life dedicated to pure thought – will be taken care of by its human partners,' predicted the astronomer Robert Jastrow in 1981, 'who will minister to its bodily needs with electricity and spare parts. Man will also provide for computer reproduction . . . We are the reproductive organs of the computer.'[91]

He was almost right. Forty years after Jastrow wrote those words, there are 2 billion computers in the world. But their numbers have been dwarfed by another more fecund automaton – the smart phone. First surpassing the world's population in 2014, there are now over 10 billion SIM cards in use worldwide, far exceeding the number of people on the planet.[92] Over 1.5 billion smart phones were sold in 2019, and they are multiplying in unprecedented numbers, outbreeding humans by a ratio of at least ten to one. Those billions of SIMs are for now largely being used by human beings. That may soon change. For while we use our devices to chat to each other across the ether, more and more of them have started talking to each other too . . .

A rich cornucopia of possibilities flowed from von Neumann's lectures on 'The General and Logical Theory of Automata' at the Hixon Symposium on Cerebral Mechanisms in Behaviour in 1948. But von Neumann was interested in more than merely proving the possibility of replicating machines. As suggested by 'cerebral mechanisms' in the symposium's title, von Neumann's automata theory also served to introduce his ideas about the workings of the brain.

Brains do not reproduce, of course – people do – and the highly cerebral von Neumann did not always carefully distinguish between the two. Still, von Neumann's automata theory and his musings about the brain both sprang from his recognition of the limits of McCulloch and Pitts's model neurons. And there is at least one sense in which the brain might be considered to be the product of a self-reproducing automaton. A brain must build itself from the bottom up without a blueprint. The several billion base-pairs-worth of data held by chromosomes inside every human cell are not sufficient to describe the brain (or any complex organ) by themselves. A neuron must interact with other neurons through some set of rules that allow the brain's incredible organization to grow from simple beginnings in the womb. That makes them in some ways kin to cellular automata, which von

Neumann, Wolfram and others have shown can produce great complexity without the help of intricate machinery or a guiding plan. At the logical level, meanwhile, von Neumann and other early computer pioneers saw their machines as primitive electronic nervous systems. Brains. Computers. Reproducing machines. It was a heady intellectual mix. A young Jeremy Bernstein, who heard von Neumann expounding his ideas as an undergraduate at Harvard University, described the talks as 'the best lectures I have ever heard on anything – like mental champagne'.[93]

In the audience of von Neumann's Hixon Symposium lecture was a recent mathematics graduate named John McCarthy. He too was electrified and decided there and then that he would try to develop thinking machines.[94] His idea was to produce smart machines by tapping evolution. 'My idea was to experiment with automata,' McCarthy says. 'One automaton interacting with another which would be its environment. You would experiment to see if you could get a smart one.' McCarthy was proposing to bootstrap his way to machine intelligence by pitting one automaton against another.

McCarthy wrote to von Neumann with the idea, and he liked it. Next year, McCarthy went to Princeton to start a doctorate and met with the great mathematician to discuss the scheme again. 'Write it up, write it up!' von Neumann urged. Despite performing a few preliminary experiments, McCarthy never did. But the passion for making machines that think, sparked by von Neumann's talk, stayed with him. It was McCarthy who later coined the term 'artificial intelligence' and, together with Marvin Minsky, founded one of the first labs dedicated to its study at MIT in the late 1950s.

Early in 1955, von Neumann was approached with an invitation to present his thoughts on the subject of computers and brains more fully by delivering the Silliman Memorial Lectures at Yale University the following year. The Silliman lectures usually spanned two weeks; but, prestigious as they were, von Neumann immediately asked if he could deliver them in just one week instead. He was a busy man: with his various research and consulting commitments for American government and military agencies, US Air Force planes were constantly being called in to jet him across the country for some important meeting or other.

But not for much longer. On 9 July 1955, von Neumann collapsed while talking on the telephone to Lewis Strauss, the AEC's chairman.[95] He was diagnosed with bone cancer the next month and was rushed to hospital for emergency surgery. By the end of the year, he was in a wheelchair, and all those important meetings were being cancelled. The one engagement he was determined to fulfil was the Silliman lectures: he worked feverishly to get his thoughts on paper and asked the Silliman committee if he could summarize them in just a day or two. But by March, it was clear that even this would be beyond him. He was admitted to hospital again, taking his manuscript with him in the hope that he might be able to edit his ideas into a form someone else could deliver on his behalf. Even that proved impossible.

Yet he had managed to flesh out his thoughts sufficiently for them to be published as a book the year after his death. In *The Computer and the Brain* von Neumann systematically compares the powers of the machines he helped to invent with the computation that goes on in the soft machine inside the human skull.[96] Brains don't look so impressive. A neuron can fire perhaps 100 times a second, while the best computers at the time were already able to carry out a million or more operations a second – and a modern laptop is at least a thousand times faster than this. Worse, neurons were billions of times less accurate than computer components: every time a signal is transmitted from one neuron to another, there is a risk that errors are exacerbated.

How, then, do brains far more mundane than von Neumann's accomplish incredible feats that defeat today's most sophisticated computers like, for example, making up an amusing pun? The answer is that neurons do not fire one after the other, but do their work simultaneously: they are not serial, like von Neumann architecture computers, but parallel – massively so. It was a lasting insight. The artificial neural networks that power today's best-performing artificial intelligence systems, like those of Google's DeepMind, are also a kind of parallel processor: they seem to 'learn' in a somewhat similar way to the human brain – altering the various weights of each artificial neuron until they can perform a particular task.

This was the first time anyone had so clearly compared brains and computers. 'Prior to von Neumann,' says inventor and futurologist Ray Kurzweil, 'the fields of computer science and neuroscience were two islands with no bridge between them.'[97]

Some believe it should have stayed that way. Psychologist Robert Epstein, for example, claims that the popularity of the 'information processing' (IP) metaphor – that is, viewing the brain as a kind of computer, as von Neumann suggested – has held back progress in neuroscience. 'The IP metaphor has had a half-century run, producing few, if any, insights along the way,' he argues. 'The time has come to hit the DELETE key.'[98] Others point out that promising alternative metaphors have been few and far between.

And the metaphor has proved extraordinarily useful for computer scientists working on neural networks and artificial intelligence – albeit after a series of false starts. As von Neumann was writing his Silliman lectures, psychologist Frank Rosenblatt was improving on the McCulloch-Pitts artificial neuron by making one that was capable of learning. Initial high hopes for this 'perceptron' faded; in 1969, Minsky and Seymour Papert published a damning book assessing the computational limitations of simple networks of artificial neurons. Funding dried up in the 1970s – the first 'AI winter' – and then renewed optimism in the 1980s gave way to a second AI winter, triggered in part by researchers who argued that progress required feedback from the senses (cameras and microphones) and interactions with the real world (through pressure or temperature sensors, for example). Despite the criticisms, in recent years, artificial intelligence algorithms have started to achieve staggering feats – from beating world champion board-game players to teaching themselves how to program – and these algorithms are often run on neural networks composed of artificial neurons not so different from Rosenblatt's perceptrons.

Some futurologists are now speculating that a superhuman artificial intelligence could transform human society beyond all recognition. That possibility has become known as the technological 'singularity' – and that term was first used by someone who had foreseen the possibility decades earlier: John von Neumann.[99]

In the eleven months that he was hospitalized, von Neumann received a stream of visitors – family, friends, collaborators and the military

'We need you.' Von Neumann receives his Medal of Freedom from President Eisenhower.

men with whom he had spent so much time in the latter years of his life. Strauss recalled 'the extraordinary picture, of sitting beside the bed of this man . . . who had been an immigrant, and there surrounding him, were the Secretary of Defense, the Deputy Secretary of Defense, the Secretaries of Air, Army, Navy, and the Chiefs of Staff'.[100] He left hospital briefly in a wheelchair to accept the Medal of Freedom from President Eisenhower. 'I wish I could be around long enough to deserve this honour,' he told the president. 'You will be with us for a long time,' Eisenhower reassured him, 'we need you.'[101]

Cancer had come at a particularly cruel time. The truth was that von Neumann had been unhappy at the IAS for several years before his death. 'Von Neumann, when I was there at Princeton, was under extreme pressure,' says Benoît Mandelbrot, who had come to the IAS in 1953 at von Neumann's invitation, 'from mathematicians, who were despising him for no longer being a mathematician; by the physicists, who were despising him for never having been a real physicist; and by everybody for having brought to Princeton this collection of low-class individuals called "programmers"'.

'Von Neumann,' Mandelbrot continues, 'was simply being shunned. And he was not a man to take it.'[102]

Von Neumann had explored several offers before settling on a free-ranging 'professor at large' position at UCLA, where he had been promised state-of-the-art computing facilities would be established. He and Klári were looking forward to a new life on the West coast but only Klári would ever move there.

Hearing that his friend was sick, Kurt Gödel wrote to von Neumann offering his condolences. 'With the greatest sorrow I have learned of your illness. The news came to me as quite unexpected,' he began. 'I hope and wish for you that your condition will soon improve even more and that the newest medical discoveries, if possible, will lead to a complete recovery.'[103]

But then Gödel, never very tactful, moves on to more important matters. 'Since you now, as I hear, are feeling stronger, I would like to allow myself to write you about a mathematical problem, of which your opinion would very much interest me . . .' He then launches into the description of a Turing machine, which if ever shown to actually exist, 'would have consequences of the greatest importance'. Namely, despite Turing's negative answer to the *Entscheidungsproblem*, Gödel said, it would imply that the discovery of some mathematical proofs can be automated. It is not known what von Neumann made of Gödel's problem or if he even saw the letter. Klári was by now answering most correspondence on his behalf. What Gödel was describing is now known as the P versus NP problem and would only be rigorously stated in 1971. Today, it is one of the most important unsolved problems in mathematics.

Marina was by now twenty-one years old and engaged to be married. Bob Whitman held a PhD in English from Harvard and had recently landed a job as an instructor at Princeton. But her father was worried that an early marriage would damage her career. 'Dear, do not misread your own character. You are very, very talented and then some,' he warned her. 'You are God's own chosen executive, and I am not joking. Besides, you like money . . . You are "genetically loaded" from both sides, both Mariette and I adore money.'

'It would be a pity, a misery to see you in petty, straightened circumstances,' he continued, 'and worst of all, cut off from using your talents and acting your proper role in life.'[104] Von Neumann eventually acquiesced to their plan with graciousness, attending their engagement party in December 1955, though he was too sick to attend the wedding itself the following year. His fears that Marina would be forced to squander her talents proved unfounded. Marina rocketed to success with the full support of her husband.

'I once asked him,' says Marina, 'when he knew he was dying, and was very upset, that "you contemplate with equanimity eliminating millions of people, yet you cannot deal with your own death." And he said, "That's entirely different."'[105]

Terrified by the prospect of his own imminent death, von Neumann asked to see the hospital's Catholic priest and returned to the faith he had ignored ever since his family had converted to it decades earlier in Budapest. 'There probably is a God,' he had once told his mother. 'Many things are easier to explain if there is than if there isn't.'

Nicholas could not bring himself to believe that his brother would 'turn overnight into a devout Catholic'. The change worried von Neumann's friends too. Ulam wrote to Strauss declaring himself 'deeply perturbed about the religious angle as it developed'. Marina says her father was thinking of Pascal's wager and had always believed that in the face of even a small possibility of suffering eternal damnation the only logical course is to be a believer before the end: 'My father told me, in so many words, once, that Catholicism was a very rough religion to live in but it was the only one to die in.'[106]

The cancer metastasized, reaching von Neumann's brain. He hallucinated in his sleep, babbling in Hungarian. Once or twice, he summoned the soldier guarding him to pass on some urgent piece of advice to the military. The mental faculties of the sharpest mind on the planet ebbed slowly away. Towards the end, says Marina, 'he asked me to test him on really simple arithmetic problems, like seven plus four, and I did this for a few minutes, and then I couldn't take it anymore; I left the room in tears.'[107]

Another frequent visitor to von Neumann's bedside was Teller. 'I have come to suspect,' he said later, 'that to most people thinking is painful. Some of us are addicted to thinking. Some of us find it a necessity. Johnny enjoyed it. I even have the suspicion that he enjoyed practically nothing else.'

'When he was dying of cancer, his brain was affected,' Teller recalled. 'I think that he suffered from this loss more than I have seen any human to suffer in any other circumstances.'[108]

Von Neumann died on 8 February 1957. He was buried in Princeton Cemetery on 12 February 1957, in a plot beside his mother, who had died of cancer just the year before, and Charles Dán, Klári's father.

There was a brief Catholic service. His friends from Los Alamos were still perplexed by his conversion. 'If Johnny is where he thought he was going,' joked one, 'there must be some very interesting conversations going on about now.'

Ulam, who probably understood the breadth of von Neumann's ideas best, lived to see many of them bloom after their creator was gone. 'He died so prematurely,' he reflected many years later, 'seeing the promised land but hardly entering it.'[109]

Klári Dán von Neumann married for the fourth and last time in 1958, this time to the American oceanographer and physicist Carl Eckart. She moved to La Jolla but could not, in the end, escape her old ghosts. Her body was found washed up on Windansea Beach on 10 November 1963. She had filled her elegant black cocktail dress with 15 pounds of wet sand and walked into the sea. The coroner's report notes that, according to her psychiatrist, she found her fourth husband 'disinterested' and 'absorbed in his work', someone who 'didn't want to go out and mix'. They slept in separate rooms at opposite ends of the house.[110]

'For the first time in my life I have relaxed and stopped chasing rainbows,' she wrote on the last page of her unfinished memoirs. The chapter entitled 'Johnny' begins, 'I would like to tell about the man, the strange contradictory and controversial person; childish and good-humoured, sophisticated and savage, brilliantly clever yet with a very limited, almost primitive lack of ability to handle his emotions – an enigma of nature that will have to remain unresolved.'[111]

Epilogue: The Man from Which Future?

'For Von Neumann the road to success was a many-laned highway with little traffic and no speed limit.'

Clay Blair Jr, 1957

'I am here to say, our house is on fire.'

Greta Thunberg, 2019

Over lunch in Los Alamos in 1950, Enrico Fermi suddenly asked his friends, 'But where is everybody?' Everyone burst out laughing. Fermi had been thumbing through a copy of the *New Yorker* and come across a cartoon blaming the recent disappearances of dustbins on extra-terrestrials. The 'Fermi Paradox' is the name now given to the conundrum of why the human race has not made contact with any alien species despite some estimates suggesting they should be legion in our galaxy. Thirty years later, Frank Tipler 'solved' this paradox.[1] Given that at least some intelligent beings would be expected to develop self-replicating machines, Tipler asks, and the billions of years such von Neumann probes would have to criss-cross the galaxy, why has there been no trace of one detected in our solar system? His conclusion is that human beings are the only intelligent species in the cosmos.

Von Neumann thought we were alone too. Shortly after Hiroshima, he had remarked semi-seriously that supernovae, the brilliant explosions caused by massive stars collapsing in on themselves, were advanced civilizations that 'having failed to solve the problem of living together, had at least succeeded in achieving togetherness by cosmic suicide'. He was keenly aware of the various ways in which his work might ultimately contribute to humanity's undoing. In coining the term 'singularity', in conversation with Ulam, von Neumann imagined a point 'in the history of the race beyond which human affairs, as we know them, could not continue'.[2] Whether that would be in a negative or positive sense remains a matter of debate: thinkers have variously speculated that an artificial superintelligence might end up fulfilling all human desires, or cosseting us like pets, or eradicating us altogether.

The cynical side of von Neumann's personality, shaped by his scrapes with totalitarianism and made famous by his transitory enthusiasm for pre-emptive war with the Soviets, often yielded to a softer face in private. 'For Johnny von Neumann I have the highest admiration in all regards,' said neurophysiologist Ralph Gerard, a contemporary of his. 'He was always gentle, always kind, always penetrating and always magnificently lucid.'[3] Shy of revealing too much of himself, his good deeds were quietly done behind people's backs. When a Hungarian-speaking factory worker in Tennessee wrote to him in 1939 asking how he could learn secondary school mathematics, von Neumann asked his friend Ortvay to send school books.[4] Benoît Mandelbrot, whose stay at the IAS had been sponsored by von Neumann, unexpectedly found himself in his debt again many years later. Sometime after von Neumann's death, prompted by a clash of personalities with his manager at IBM, Mandelbrot went looking for a new job – and found that the way had been made easier for him. Von Neumann had spread the word widely that his research could be of great significance – but was very risky. 'He may really sink,' von Neumann warned them, says Mandelbrot. 'If he's in trouble, please help him.'[5]

Which of these was the real von Neumann? 'Both were real,' Marina says.[6] But the dissonance between them confused even her, she admits. Beneath the surface the two facets of his personality were at war. Von Neumann hoped the best in people would triumph and tried to be as magnanimous and honourable as possible. But experience and reason taught him to avoid placing too much faith in human virtue.

Nowhere is the tug-of-war between the cool rationalist and kind

Courtesy of Marina von Neumann Whitman.

philanthropist more evident than in von Neumann's remarkable meditation on the existential threats facing humanity in the decades to come. Published in June 1955 in *Fortune* magazine, 'Can We Survive Technology?' begins with a dire warning: 'literally and figuratively, we are running out of room'.[7] Advances in domains such as weaponry and telecommunications have greatly increased the speed with which conflicts can escalate and magnified their scope. Regional disputes can quickly engulf the whole planet. 'At long last,' he continues, 'we begin to feel the effects of the finite, actual size of the earth in a critical way.'

Long before climate change became a widely discussed concern, the essay shows von Neumann was alert to the idea that carbon dioxide emissions from burning coal and oil were warming the planet. He favoured the idea of coming up with new geo-engineering technologies to control the climate by, for example, painting surfaces to change how much sunlight they reflect – quite likely the first time that anyone had talked about deliberately warming or cooling the earth in this way. Interventions such as these, he predicts, 'will merge each nation's affairs with those of every other, more thoroughly than the threat of a nuclear or any other war may already have done.'

Von Neumann speculates that nuclear reactors will rapidly become more efficient and held out hope that mankind would harness fusion too in the long term. Automation would continue, he predicted, accelerated by advances in solid-state electronics that will bring much faster computing machines. But all technological progress, he warns, will also inevitably be harnessed for military use. Sophisticated forms of climate control, for example, could 'lend themselves to forms of climatic warfare as yet unimagined'.

Preventing disaster will require the invention of 'new political forms and procedures' (and the Intergovernmental Panel on Climate Change, established in 1988, arguably embodies one attempt to do exactly that). But what we cannot do, he says, is stop the march of ideas. 'The very techniques that create the dangers and the instabilities are in themselves useful, or closely related to the useful,' he argues. Under the ominous heading 'Survival – A Possibility', he continues: 'For progress there is no cure. Any attempt to find automatically safe channels for the present explosive variety of progress must lead to frustration. The only safety possible is relative, and it lies in an intelligent exercise of day-to-day judgment.'

There is, as he puts it, no 'complete recipe' – no panacea – for avoiding extinction at the hands of technology. 'We can specify only the human qualities required: patience, flexibility, intelligence.'

Select Bibliography

Abella, Alex, 2008, *Soldiers of Reason: The RAND Corporation and the Rise of the American Empire,* Harcourt, San Diego, Calif.

Aspray, William, 1990, *John von Neumann and the Origins of Modern Computing*, MIT Press, Cambridge, Mass.

Baggott, Jim, 2003, *Beyond Measure: Modern Physics, Philosophy and the Meaning of Quantum Theory*, Oxford University Press, Oxford.

Baggott, Jim, 2009, *Atomic: The First War of Physics and the Secret History of the Atom Bomb: 1939–49*, Icon Books, London.

Binmore, Ken, 2007, *Game Theory: A Very Short Introduction*, Oxford University Press, Oxford.

Burks, Arthur W., 1966, *Theory of Self-reproducing Automata*, University of Illinois Press, Urbana.

Byrne, Peter, 2010, *The Many Worlds of Hugh Everett III: Multiple Universes, Mutual Assured Destruction, and the Meltdown of a Nuclear Family*, Oxford University Press, Oxford.

Copeland, Jack B. (ed.), 2004, *The Essential Turing: Seminal Writings in Computing, Logic, Philosophy, Artificial Intelligence, and Artificial Life Plus The Secrets of Enigma*, Oxford University Press, Oxford.

Davis, Martin, 2000, *The Universal Computer: The Road from Leibniz to Turing*, W. W. Norton & Company, New York.

Dawson, John W., 1997, *Logical Dilemmas: The Life and Work of Kurt Gödel,* A. K. Peters, Wellesley, Mass.

Drexler, Eric, 1986, *Engines of Creation*, Doubleday, New York.

Dyson, Freeman, 1979, *Disturbing the Universe*, Harper and Row, New York.

Dyson, George, 2012, *Turing's Cathedral*, Pantheon Books, New York.

Einstein, Albert, 1922, *Sidelights on Relativity*, E. P. Dutton and Company, New York.

Erickson, Paul, 2015, *The World the Game Theorists Made*, The University of Chicago Press, Chicago.

Frank, Tibor, 2007, *The Social Construction of Hungarian Genius (1867–1930)*, Eötvös Loránd University, Budapest.

Freitas, Robert A., Jr and Merkle, Ralph C., 2004, *Kinematic Self-Replicating Machines*, Landes Bioscience, Georgetown, Tex., http://www.MolecularAssembler.com/KSRM.htm.

Goldstein, Rebecca, 2005, *Incompleteness: The Proof and Paradox of Kurt Gödel*, W. W. Norton & Company, New York.

Goldstine, Herman H., 1972, *The Computer from Pascal to von Neumann*, Princeton University Press, Princeton.

Gowers, Timothy (ed.), 2008, *The Princeton Companion to Mathematics*, Princeton University Press, Princeton.

Gray, Jeremy, 2000, *The Hilbert Challenge*, Oxford University Press, Oxford.

Gray, Jeremy, 2008, *Plato's Ghost: The Modernist Transformation of Mathematics*, Princeton University Press, Princeton.

Haigh, Thomas, Priestley, Mark and Rope, Crispin, 2016, *ENIAC in Action: Making and Remaking the Modern Computer*, MIT Press, Cambridge, Mass.

Hargittai, István, 2006, *The Martians of Science: Five Physicists Who Changed the Twentieth Century*, Oxford University Press, Oxford.

Heims, Steve J., 1982, *John von Neumann and Norbert Wiener: From Mathematics to the Technologies of Life and Death*, MIT Press, Cambridge, Mass.

Hoddeson, Lillian, Henriksen, Paul W., Meade, Roger A. and Westfall, Catherine, 1993, *Critical Assembly: A Technical History of Los Alamos during the Oppenheimer Years, 1943–1945*, Cambridge University Press, Cambridge.

Hodges, Andrew, 2012, *Alan Turing: The Enigma. The Centenary Edition*, Princeton University Press, Princeton.

Jammer, Max, 1974, *The Philosophy of Quantum Mechanics: The Interpretations of Quantum Mechanics in Historical Perspective*, Wiley, Hoboken.

Jardini, David, 2013, *Thinking Through the Cold War: RAND, National Security and Domestic Policy, 1945–1975*, Smashwords.

Kaplan, Fred, 1983, *The Wizards of Armageddon*, Stanford University Press, Stanford.

Kármán, Theodore von with Edson, Lee, 1967, *The Wind and Beyond: Theodore von Kármán, Pioneer in Aviation and Pathfinder in Space*, Little, Brown and Co., Boston.

Leonard, Robert, 2010, *Von Neumann, Morgenstern, and the Creation of Game Theory: From Chess to Social Science, 1900–1960*, Cambridge University Press, Cambridge.

Levy, Steven, 1993, *Artificial Life: A Report from the Frontier Where Computers Meet Biology*, Vintage, New York.

Lukacs, John, 1998, *Budapest 1900: A Historical Portrait of a City and Its Culture*, Grove Press, New York.

Macrae, Norman, 1992, *John von Neumann: The Scientific Genius Who Pioneered the Modern Computer, Game Theory, Nuclear Deterrence and Much More*, Pantheon Books, New York.

McDonald, John, 1950, *Strategy in Poker, Business and War*, W. W. Norton, New York.

Musil, Robert, 1931–3, *Der Mann ohne Eigenschaften*, Rowohlt Verlag, Berlin, English edition: 1997, *The Man without Qualities*, trans. Sophie Wilkins, Picador, London.

Nasar, Sylvia, 1998, *A Beautiful Mind*, Simon & Schuster, New York.

Neumann, John von, 2005, *John von Neumann: Selected Letters*, ed. Miklós Rédei, American Mathematical Society, Providence, R.I.

Neumann, John von, 2012, *The Computer and the Brain*, Yale University Press, New Haven (first published 1958).

Neumann, John von, 2018, *Mathematical Foundations of Quantum Mechanics*, Princeton University Press, Princeton.

Neumann, John von and Morgenstern, Oskar, 1944, *Theory of Games and Economic Behavior*, Princeton University Press, Princeton.

Petzold, Charles, 2008, *The Annotated Turing: A Guided Tour Through Alan Turing's Historic Paper on Computability and the Turing Machine*, Wiley, Hoboken.

Poundstone, William, 1992, *Prisoner's Dilemma: John von Neumann, Game Theory and the Puzzle of the Bomb*, Doubleday, New York.

Reid, Constance, 1986, *Hilbert-Courant*, Springer, New York.

Sime, Ruth Lewin, 1996, *Lise Meitner: A Life in Physics*, University of California Press, Berkeley.

Susskind, Leonard, 2015, *Quantum Mechanics: The Theoretical Minimum*, Penguin, London.

Taub, A. H. (ed.), 1963, *Collected Works of John von Neumann*, 6 volumes, New York: Pergamon Press.

Teller, Edward (with Judith Shoolery), 2001, *Memoirs: A Twentieth-Century Journey in Science and Politics*, Perseus, Cambridge, Mass.

Ulam, Stanisław M. 1991, *Adventures of a Mathematician*, University of California Press, Berkeley.

Von Kármán, Theodore with Edson, Lee, 1967, *The Wind and Beyond: Theodore von Kármán, Pioneer in Aviation and Pathfinder in Space*, Little, Brown and Co., Boston.

Von Neumann, John, 2005, *John von Neumann: Selected Letters*, ed. Miklós Rédei, American Mathematical Society, Providence, R.I.

Von Neumann, John, 2012, *The Computer and the Brain*, Yale University Press, New Haven (first published 1958).

Von Neumann, John, 2018, *Mathematical Foundations of Quantum Mechanics*, Princeton University Press, Princeton.

Von Neumann, John, and Morgenstern, Oskar, 1944, *Theory of Games and Economic Behavior*, Princeton University Press, Princeton.

Vonneuman, Nicholas A., 1987, *John von Neumann as Seen by His Brother*, P.O. Box 3097 Meadowbrook, Pa.

Whitman, Marina von Neumann, 2012, *The Martian's Daughter*, University of Michigan Press, Ann Arbor.

Wolfram, Steven, 2002, *A New Kind of Science*, Wolfram Media, Champagne, Ill.

Notes

INTRODUCTION: WHO WAS JOHN VON NEUMANN?

1. Albert Einstein, 1922, *Sidelights on Relativity*, E. P. Dutton and Company, New York.
2. Freeman Dyson, 2018, personal communication.

CHAPTER 1: MADE IN BUDAPEST

1. Renamed Bajcsy-Zsilinsky Street in 1945 after a resistance hero.
2. John Lukacs, 1998, *Budapest 1900: A Historical Portrait of a City and Its Culture*, Grove Press, New York.
3. Robert Musil, 1931–3, *Der Mann ohne Eigenschaften*, Rowohlt Verlag, Berlin. English edition: 1997, *The Man without Qualities*, trans. Sophie Wilkins, Picador, London.
4. Nicholas A. Vonneuman, 1987, *John von Neumann as Seen by His Brother*, P.O. Box 3097, Meadowbrook, Pa.
5. Ibid.
6. Formidable but not infallible. His school friend Eugene Wigner recalled that when Johnny had once tried to multiply two five-digit numbers in his head, he got the answer wrong. 'Then why on earth are you congratulating me?' a bewildered Jancsi had asked him. Because, Wigner said, Johnny had been very close.
7. See, for example, Harry Henderson, 2007, *Mathematics: Powerful Patterns into Nature and Society*, Chelsea House, New York, p. 30.
8. Despite decades of research, whether chess ability is correlated with general intelligence or mathematical ability is still a fiercely contested question. A recent meta-analysis of past studies (Alexander P. Burgoyne et al., 'The Relationship between Cognitive Ability and Chess Skill: A Comprehensive Meta-analysis', *Intelligence*, 59 (2016), pp. 72–83)

suggests some correlation exists and is strongest for numerical ability, and among younger players rather than highly skilled older masters of the game. One of the earliest studies in the area found no differences in intelligence between eight grandmasters and non-chess players (I. N. Djakow, N. W. Petrowski and P. A. Rudik, 1927, *Psychologie des Schachspiels*, deGruyter, Berlin).

9. Klára von Neumann, *Johnny*, quoted in George Dyson, 2012, *Turing's Cathedral*, Pantheon Books, New York.
10. Vonneuman, *John von Neumann as Seen by His Brother*.
11. Ibid.
12. Ibid. Von Neumann was among several early computer pioneers to be inspired by this ingenious mechanism, which allowed fabrics with complex patterns to be mass-produced for the first time. Like the loom, the Analytical Engine, a mechanical computer designed in the 1830s by the English polymath Charles Babbage, was to be programmed with strings of punched cards. 'We may say most aptly that the Analytical Engine weaves algebraic patterns just as the Jacquard loom weaves flowers and leaves,' wrote Babbage's collaborator, the mathematician Countess Ada Lovelace. Sadly, Babbage never managed to raise the sizeable sum he needed to build the machine. Jacquard's place in the history of computing was assured by his loom's influence on Herman Hollerith, an enterprising civil servant who worked for the United States Census Bureau. Less than twenty years after Babbage's death in 1871, Hollerith was granted a patent for an electromechanical device 'to simplify and thereby facilitate the compilation of statistics' by recording information using holes in a roll of paper tape – though he quickly switched to more robust punched cards. Hollerith's machines were used for the 1890 American census, and the improvements in efficiency they brought about led dozens of other countries to lease his equipment. In 1911, his company and three others were amalgamated to form what became IBM, a firm that von Neumann would influence when he was older, encouraging them to switch from mechanical computers to electronic ones.
13. Of Johnny's brothers, Nicholas merged the 'von' with 'Neumann' after he arrived in the United States, becoming Vonneuman. Michael simply used Neumann.
14. Theodore von Kármán with Lee Edson, 1967, *The Wind and Beyond: Theodore von Kármán, Pioneer in Aviation and Pathfinder in Space*, Little, Brown and Co., Boston.
15. Tibor Frank, 2007, *The Social Construction of Hungarian Genius (1867–1930)*, Eötvös Loránd University, Budapest.

16. George Klein, 1992, *The Atheist and the Holy City: Encounters and Reflections*, MIT Press, Cambridge, Mass.
17. Edward Teller (with Judith Shoolery), 2001, *Memoirs: A Twentieth-Century Journey in Science and Politics*, Perseus, Cambridge, Mass.
18. Stanisław M. Ulam, 1991, *Adventures of a Mathematician*, University of California Press, Berkeley.
19. Science rarely lives up to this ideal in the real world.

CHAPTER 2: TO INFINITY AND BEYOND

1. After being forced out of his job at the University of Königsberg by the Nazis in 1933.
2. George Pólya Papers, quoted in Frank, *The Social Construction of Hungarian Genius*.
3. Gabor Szegő Papers, quoted in ibid.
4. M. Fekete and J. L. von Neumann, 'Über die Lage der Nullstellen gewisser Minimumpolynome', *Jahresbericht der Deutschen Mathematiker-Vereinigung*, 31 (1922).
5. Timothy Gowers (ed.), 2008, *The Princeton Companion to Mathematics*, Princeton University Press, Princeton.
6. Freeman Dyson, 'A Walk through Johnny von Neumann's Garden', *Notices of the American Mathematical Society*, 60(2) (2013), pp. 154–61.
7. Andrew Janos, 1982, *The Politics of Backwardness in Hungary*, Princeton University Press, Princeton.
8. Vonneuman, *John von Neumann as Seen by His Brother*.
9. Hearings (1955), United States: US Government Printing Office.
10. Pál Prónay, 1963, *A hatarban a halal kaszal: Fejezetek Prónay Pal feljegyzeseibol*, ed. Agnes Szabo and Ervin Pamlenyi, Kossuth Könyvkiadó, Budapest.
11. Eugene P. Wigner, 'Two Kinds of Reality', *The Monist*, 48(2) (1964), pp. 248–64.
12. Quoted in the definitive work on mathematical modernism: Jeremy Gray, 2008, *Plato's Ghost: The Modernist Transformation of Mathematics*, Princeton University Press, Princeton.
13. Ibid.
14. P. Stäckel, 1913, *Wolfgang und Johann Bolyai: Geometrische Untersuchungen, Leben und Schriften der beiden Bolyai*, Teubner, Leipzig, quoted in ibid.

15. Much of what follows is based on the accounts in Constance Reid, 1986, *Hilbert-Courant*, Springer, New York; and Gray, *Plato's Ghost.*
16. Reid, *Hilbert-Courant.*
17. Bertrand Russell, 1967, *The Autobiography of Bertrand Russell: 1872–1914* (2000 edn), Routledge, New York.
18. Ibid.
19. Reid, *Hilbert-Courant.*
20. John von Neumann, 'Zur Einführung der transfiniten Zahlen', *Acta Scientiarum Mathematicarum (Szeged)*, 1(4) (1923), pp. 199–208.
21. Von Kármán, *The Wind and Beyond.* In 2008, the world would learn that, to the contrary, mathematics can make lots of money – and lose it too.
22. Wigner's dream of pursuing theoretical physics after school had ended less happily than von Neumann's. 'And exactly how many jobs are there in Hungary for theoretical physicists?' his father asked him. 'Four,' Wigner replied. He was promptly packed off to study chemical engineering in Berlin. Fellner would initially study the subject for similar reasons. All three would drop it quite soon after finishing their degrees to pursue their true passions.
23. Quoted in Stanisław Ulam, 'John von Neumann 1903–1957', *Bulletin of the American Mathematical Society*, 64 (1958), pp. 1–49.
24. John von Neumann, 'Eine Axiomatisierung der Mengenlehre', *Journal für die reine und angewandte Mathematik*, 154 (1925), pp. 219–40.
25. John von Neumann, 'Die Axiomatisierung der Mengenlehre', *Mathematische Zeitschrift*, 27 (1928), pp. 669–752.
26. Quoted in Dyson, *Turing's Cathedral.*

CHAPTER 3: THE QUANTUM EVANGELIST

1. Einstein to Heinrich Zangger, 12 May 1912, *The Collected Papers of Albert Einstein*, Princeton University Press, Princeton, vol. 5, p. 299, quoted in Manjit Kumar, 2008, *Quantum: Einstein, Bohr and the Great Debate about the Nature of Reality*, Icon Books, London.
2. Werner Heisenberg, 'Über quantentheoretische Umdeutung kinematischer und mechanischer Beziehungen', *Zeitschrift für Physik*, 33(1) (1925), pp. 879–93.
3. Well, not quite instantaneously it turns out. The results of some remarkable experiments published in June 2019 show transitions take

a small but finite amount of time (many microseconds): https://doi.org/10.1038/s41586-019-1287-z.

4. An electron in the energy state n = 6 might, for example, drop down to the ground state or arrive there via n = 2, effectively making two instantaneous mini transitions; one from n = 6 to n = 2, and a second from n=2 to the ground state.
5. If I roll a dice twice and want to know the probability that I roll a 3 (1 in 6 or 1/6) followed by a 4 (ditto), then I multiply the two probabilities together; the answer is 1 in 36 or 1/36.
6. Here is one of Heisenberg's three-by-three arrays multiplied by itself. The first digit in the resulting array is a result of calculating 2×2+5×1+4×4 = 25; the second digit of the first row is 2×5+5×1+4×2 = 23 and so on.

$$\begin{matrix} 2 & 5 & 4 \\ 1 & 1 & 3 \\ 4 & 2 & 7 \end{matrix} \times \begin{matrix} 2 & 5 & 4 \\ 1 & 1 & 3 \\ 4 & 2 & 7 \end{matrix} = \begin{pmatrix} 25 & 23 & 51 \\ 15 & 12 & 28 \\ 38 & 36 & 71 \end{pmatrix}$$

Jeremy Bernstein explains this superbly in more detail in his paper 'Max Born and the Quantum Theory', *American Journal of Physics*, 73 (2005), pp. 999–1008.

7. For example, $\begin{matrix} 1 & 3 \\ 4 & 2 \end{matrix} \times \begin{matrix} 2 & 5 \\ 1 & 3 \end{matrix} = \begin{matrix} 5 & 14 \\ 10 & 26 \end{matrix}$. Flip the order of multiplication and you get $\begin{matrix} 22 & 16 \\ 13 & 9 \end{matrix}$.
8. Kumar, *Quantum*.
9. Born to Einstein, 15 July 1925, in Max Born, 2005, *The Born–Einstein Letters 1916–1955: Friendship, Politics and Physics in Uncertain Times*, Macmillan, New York.
10. Born's equation connecting the position of a particle, x, with its momentum, p:

$$xp - px = i\frac{h}{2\pi}I$$

On the right-hand side is h, Planck's constant; I, the identity or unit matrix; and the imaginary number i (defined by the property that its square is −1). Unit matrices have 1s on their diagonal and 0s everywhere else, like this $\begin{matrix} 1 & 0 \\ 0 & 1 \end{matrix}$. The I in the equation allows both sides to be written as matrices. Imaginary numbers obey the same mathematical rules as real numbers and have proved incredibly useful in physics and engineering (for example in the very non-imaginary realm of circuit theory). In quantum mechanics, they are a convenience that makes the equations easier to solve. They make the maths work.

11. The uncertainty principle states that if Δx and Δp are the respective uncertainties in the measurements of position and momentum, their product will always be at least $\frac{h}{4\pi}$. That is, $x\ p = \frac{h}{4\pi}$. The more accurately we know the position of a particle, the *less* accurately can we know its momentum. Physics students are still sometimes taught that this uncertainty arises because of disturbances introduced by the act of measurement. This is not true. While in practice the measuring process does introduce some additional uncertainty, Heisenberg's uncertainty principle places a fundamental limit on the precision with which a particle's position and momentum can ever be determined, no matter how delicately they are measured. The uncertainty is an inherent property of the particle in question.
12. Louis de Broglie, 'XXXV. A Tentative Theory of Light Quanta', *The London, Edinburgh, and Dublin Philosophical Magazine and Journal of Science*, 47(278) (1924), pp. 446–58.
13. Johnny's compatriot Leo Szilard, recovered by now from his 'intolerably boring' maths lessons in Budapest, would try to patent the idea of using electron waves to image small objects, and Ernst Ruska and Max Knoll would (independently) build the first prototype of an electron microscope in 1931.
14. If you're wondering how a function can satisfy an equation, a relatively simple example is $f(x) + f(y) = x + y$, which is satisfied by, for example, the function $f(x) = x$.
15. $\psi = a_1\psi_1 + a_2\psi_2 + a_3\psi_3 + a_4\psi_4 + \ldots$ where a_n is the fraction each wave function makes up of the total.
16. Kumar, *Quantum*, p. 225.
17. Ibid.
18. Imagine writing a list of numbers beginning 1, 2, 3 ... One can imagine writing next to each number an element from a Heisenberg matrix.
19. A bounded particle, such as the electron of a hydrogen atom, can also be in an infinite superposition of states. But, as we have seen, these states are countable: it is possible to assign quantum numbers to them. But a physical theory like quantum mechanics has to be able to cope with unbounded particles whizzing freely through space too.

20. The coordinates x, y, z are real numbers, so can be any point along a number line (the sort used to teach primary school kids how to add up). The real numbers include negative numbers, fractions and the irrational numbers, like pi and the square root of 2, which can't be expressed as a fraction (they go on for ever after the decimal point, without repeating).
21. John von Neumann, 2018, *Mathematical Foundations of Quantum Mechanics*, Princeton University Press, Princeton.
22. Ian McEwan, 2010, *Solar*, Random House, London.
23. Freeman Dyson, quoted in Graham Farmelo, 2009, *The Strangest Man: The Hidden Life of Paul Dirac, Quantum Genius*, Faber and Faber, London.
24. Paul A. M. Dirac, 'The Fundamental Equations of Quantum Mechanics', *Proceedings of the Royal Society of London. Series A, Containing Papers of a Mathematical and Physical Character*, 109(752) (1925), pp. 642–53.
25. Paul Dirac, 1930, *The Principles of Quantum Mechanics*, Oxford University Press, Oxford.
26. Dyson, 'A Walk through Johnny von Neumann's Garden', p. 154.
27. Operators are expressed as partial derivatives in wave mechanics.
28. Max Jammer, 1974, *The Philosophy of Quantum Mechanics: The Interpretations of Quantum Mechanics in Historical Perspective*, Wiley, Hoboken.
29. So $x_1^2+x_2^2+x_3^2+x_4^2+x_5^2$ . . . etc. has to be less than infinity. Mathematicians say the series of numbers has to 'converge'.
30. What does a Hilbert space of quantum states look like? As the magnitude of each wave function (squared) is 1, their vectors are all of length one. If Hilbert space was merely two-dimensional, the collection of all possible state vectors would describe a circle centred on the origin, and in three dimensions they would describe the surface of a sphere. Since Hilbert space is infinite-dimensional, the tips of the state vectors representing the wave functions touch the surface of an infinite-dimensional ball, known as a hypersphere.
31. For those in the know, the orthogonal functions in question are a Fourier series of sines and cosines.
32. The Chebyshev polynomials of von Neumann's very first paper are orthogonal.
33. Like this: $\psi = c_1f_1+c_2f_2+c_3f_3+c_4f_4+$ . . . Here, the *f*s are the orthogonal functions and the *c*s the coefficients.
34. I.e. $|c_1|^2+|c_2|^2+|c_3|^2+|c_4|^2+$ . . . etc. $= 1$.
35. Von Neumann, *Mathematical Foundations of Quantum Mechanics*.

36. Frank, *The Social Construction of Hungarian Genius*.
37. Interview of Eugene Wigner by Charles Weiner and Jagdish Mehra, 30 November 1966, Niels Bohr Library and Archives, American Institute of Physics, College Park, MD USA, www.aip.org/history-programs/niels-bohr-library/oral-histories/4964.
38. 'With Nordheim and (nominally) Hilbert': David Hilbert, John von Neumann and Lothar Nordheim, 'Über die Grundlagen der Quantenmechanik', *Mathematische Annalen*, 98 (1927), pp. 1–30; 'later on his own': J. von Neumann, 'Mathematische Begründung der Quantenmechanik', *Nachrichten von der Gesellschaft der Wissenschaften zu Göttingen* (1927), pp. 1–57, and John von Neumann, 'Allgemeine Eigenwerttheorie Hermitescher Funktionaloperatoren', *Mathematische Annalen*, 102 (1929), pp. 49–131.
39. Physicist Erich Hückel is credited with writing the poem. Translated by Felix Bloch. See Elisabeth Oakes, 2000, *Encyclopedia of World Scientists*, Facts on File, New York.
40. Steven Weinberg, 'The Trouble with Quantum Mechanics', *The New York Review of Books*, 19 January 2017.
41. Niels Bohr, 'Wirkungsquantum und Naturbeschreibung', *Naturwiss*, 17 (1929), pp. 483–6. First translated into English in 1934 as *The Quantum of Action and the Description of Nature*, Cambridge University Press, Cambridge.
42. Einstein to Max Born, 3 March 1947, in Born, *The Born–Einstein Letters 1916–1955*.
43. In standard histories of quantum mechanics, the Copenhagen interpretation first emerges in 1927, after being presented by Bohr at the fifth Solvay Conference in Brussels. The story goes that Bohr's influential acolytes then promoted Bohr's views until they became orthodoxy. The truth, however, appears to be that Bohr's views are not perfectly aligned with the Copenhagen interpretation, and his writings were not widely read at the time. In fact, the phrase 'Copenhagen interpretation' first turns up only in 1955, in an essay by Heisenberg, who appears to have set down the ideas that scientists often regard as the canonical view of quantum mechanics (see, for example, Don Howard, 'Who Invented the "Copenhagen Interpretation?" A Study in Mythology', *Philosophy of Science*, 71(5) (2004), pp. 669–82, doi:10.1086/425941). Nonetheless, different aspects of the Copenhagen interpretation appeared long before Heisenberg collected them together under a single banner, and von Neumann's work on the measurement problem is one of the earliest contributions to that synthesis.

44. David N. Mermin, 'Could Feynman Have Said This?', *Physics Today*, 57(5) (2004), pp 10–12.
45. Wigner changed his mind later.
46. Abraham Pais, 'Einstein and the Quantum Theory', *Reviews of Modern Physics*, 51 (1979), pp. 863–914.
47. *Mathematical Foundations of Quantum Mechanics* was not translated into English until 1955.
48. Andrew Hodges, 2012, *Alan Turing: The Enigma. The Centenary Edition*, Princeton University Press, Princeton.
49. Erwin Schrödinger, 'Die gegenwärtige Situation in der Quantenmechanik', *Naturwissenschaften*, 23(48) (1935), pp. 807–12.
50. Einstein to Born, 4 December 1926, in Born, *The Born–Einstein Letters 1916–1955*.
51. If they *were* measurable, then quantum mechanics could easily be proved wrong
52. This sounds like a fudge but need not be. There are many examples of hidden variable-like theories in physics that have turned out to be useful. The ideal gas law, for example, relates the pressure, volume and temperature of a fixed amount of gas. But there is a deeper 'hidden variables' theory – the kinetic theory of gases – that can be used to derive the law by considering the behaviour of the atoms or molecules of the gas bouncing around inside a vessel. Put another way, the ideal gas law is a result of the 'hidden' motions of gas particles.
53. Jammer, *The Philosophy of Quantum Mechanics*.
54. Andrew Szanton, 1992, *The Recollections of Eugene P. Wigner: As Told to Andrew Szanton*, Springer, Berlin.
55. Accounts differ on this. Wigner says the von Neumanns arrived a day after him. Von Neumann's biographer claims they arrived a week later (Norman Macrae, 1992, *John von Neumann: The Scientific Genius Who Pioneered the Modern Computer, Game Theory, Nuclear Deterrence and Much More*, Pantheon Books, New York).
56. David N. Mermin, 'Hidden Variables and the Two Theorems of John Bell', *Reviews of Modern Physics*, 65 (1993), pp. 803–15.
57. This and much of what follows is from Elise Crull and Guido Bacciagaluppi (eds.), 2016, *Grete Hermann: Between Physics and Philosophy*, Springer, Berlin, Heidelberg, New York.
58. Werner Heisenberg, 1971, *Physics and Beyond: Encounters and Conversations*, Harper and Row, New York.
59. The additivity postulate states that the sum of the expectation (average) values of two operators (acting on a system) is equal to the

expectation value of their sum. This is true in both quantum and classical physics. For example, the average energy of a particle (the sum of the kinetic and potential energy of the particle) is equal to its average kinetic energy plus its average potential energy).

60. The essay was never published but has recently been recovered from Dirac's archives and translated into English. See Crull and Bacciagaluppi, *Grete Hermann*.
61. Grete Hermann, 'Die naturphilosophischen Grundlagen der Quantenmechanik', *Abhandlugen der Fries'schen Schule*, 6(2) (1935), pp. 75–152.
62. Grete Hermann, 'Die naturphilosophischen Grundlagen der Quantenmechanik', *Die Naturwissenschaften*, 23(42) (1935), pp. 718–21.
63. The philosophy Hermann eventually expounded was a version of what is now known as the relational interpretation of quantum mechanics. Carlo Rovelli is generally credited with first describing the idea in 1994. Hermann got to it over half a century earlier. In relational interpretations, roughly speaking, quantum theory only describes the state of a system relative to other systems or observers. There is no 'objective' observer-independent state of a system 'out there'. Two observers may have differing views on a quantum event until they compare them. According to Hermann, any observer can reconstruct the sequence of events that led to a particular observation, so causality, of sorts, is restored. This is all the causality required for most Kantians.
64. John Stewart Bell, interview in *Omni*, May 1988.
65. Quoted in Nicholas Gisin, 'Sundays in a Quantum Engineer's Life' (2001), arXiv:quant-ph/0104140, https://arxiv.org/abs/quant-ph/0104140.
66. John Stewart Bell, 1987, *Speakable and Unspeakable in Quantum Mechanics*, Cambridge University Press, Cambridge.
67. From 'On the Impossible Pilot Wave', republished in ibid.
68. John Stewart Bell, 1966, 'On the Problem of Hidden Variables in Quantum Mechanics', *Reviews of Modern Physics*, 38 (1966), pp. 447–52.
69. N. D. Mermin, 'Hidden Variables and the Two Theorems of John Bell', *Reviews of Modern Physics*, 65 (1993), pp. 803–15.
70. Jeffrey Bub, 'Von Neumann's "No Hidden Variables" Proof: A Re-Appraisal', *Foundations of Physics*, 40 (2010), pp. 1333–40; D. Dieks, 'Von Neumann's Impossibility Proof: Mathematics in the Service of Rhetorics', *Studies in History and Philosophy of Modern Physics*, 60 (2017), pp. 136–48.

71. Michael Stöltzner, 1999, 'What John von Neumann Thought of the Bohm Interpretation', in D. Greenberger et al. (eds.), *Epistemological and Experimental Perspectives on Quantum Physics*, Kluwer Academic Publishers, Dordrecht.
72. Einstein to Born, 12 May 1952, in Born, *The Born–Einstein Letters 1916–1955*.
73. Albert Einstein, Boris Podolsky and Nathan Rosen, 1935, 'Can Quantum-Mechanical Description of Physical Reality Be Considered Complete?', *Physical Review*, 47(10) (1935), pp. 777–80.
74. Using entangled photons rather than hydrogen atoms is easier for such experiments.
75. This and what follows is from Stefano Osnaghi, Fábio Freitas and Olival Freire Jr, 'The Origin of the Everettian Heresy', *Studies in History and Philosophy of Modern Physics*, 40, pp. 97–123.
76. Peter Byrne, 2010, *The Many Worlds of Hugh Everett III: Multiple Universes, Mutual Assured Destruction, and the Meltdown of a Nuclear Family*, Oxford University Press, Oxford.
77. Philip Ball discusses the problems and advantages of this and other interpretations of quantum mechanics in his wonderful 2018 book *Beyond Weird*, The Bodley Head, London.
78. G. C. Ghirardi, A. Rimini and T. Weber (1986), 'Unified Dynamics for Microscopic and Macroscopic Systems', *Physical Review D*, 34(2) (1986), pp. 470–91.
79. Von Neumann, *Mathematical Foundations of Quantum Mechanics*.
80. P. A. M. Dirac, 1978, *Directions in Physics*, Wiley, New York.
81. Laurent-Moïse Schwartz tidied up the mathematics of Dirac's 'improper' delta functions in 1945 and was awarded the prestigious Fields Medal for his efforts.
82. Jammer, *The Philosophy of Quantum Mechanics*.
83. Many thanks to Ulrich Pennig, without whom I would have nothing at all to say about von Neumann algebras.
84. Dyson, 'A Walk through Johnny von Neumann's Garden'.
85. Carlo Rovelli, 2018, *The Order of Time*, Allen Lane, London.
86. Von Neumann to O. Veblen, 19 June 1933, Library of Congress archives.
87. Fabian Waldinger, 'Bombs, Brains, and Science: The Role of Human and Physical Capital for the Creation of Scientific Knowledge', *Review of Economics and Statistics*, 98(5) (2016), pp. 811–31.
88. A. Fraenkel, 1967, *Lebenskreise*, translated and quoted by David. E. Rowe, 1986, '"Jewish Mathematics" at Gottingen in the Era of Felix Klein', *Isis*, 77(3), pp. 422–49.

CHAPTER 4: PROJECT Y AND THE SUPER

1. Paul Halmos, 'The Legend of John von Neumann', *The American Mathematical Monthly*, 80(4) (1973), pp. 382–94.
2. Mariette would later be instrumental in establishing Brookhaven National Laboratory on Long Island, working there for some twenty-eight years as a senior administrator. See Marina von Neumann Whitman, *The Martian's Daughter*, University of Michigan Press, Ann Arbor, and https://www.bnl.gov/60th/EarlyBNLers.asp.
3. Richard Feynman with Ralph Leighton, 1985, *Surely You're Joking, Mr. Feynman!: Adventures of a Curious Character*, W. W. Norton, New York.
4. Ergodic theory crops up in many diverse areas of physics and mathematics. Mathematicians Terence Tao and Ben Green used the theorem in 2004 to prove a conjecture related to prime numbers dating from the 1770s. They showed that there are sequences of prime numbers of arbitrary length where the difference between adjacent terms is a constant. For instance, 3,5,7 is one such sequence of length three.
5. The dispute between the two is discussed in Joseph D. Zund, 'George David Birkhoff and John von Neumann: A Question of Priority and the Ergodic Theorems, 1931–1932', *Historia Mathematica*, 29 (2002), pp. 138–56.
6. Garrett Birkhoff, 1958, 'Von Neumann and Lattice Theory', *Bulletin of the American Mathematical Society*, 64 (1958), pp. 50–56.
7. Alan Turing, 'On Computable Numbers, with an Application to the *Entscheidungsproblem*', published in two parts 1936–7, *Proceedings of the London Mathematical Society*, 42(1) (1937), pp. 230–65.
8. Interview with Herman Goldstine conducted by Nancy Stern, 1980, https://conservancy.umn.edu/bitstream/handle/11299/107333/oho18hhg.pdf?sequence=1&isAllowed=y.
9. Ulam, *Adventures of a Mathematician*.
10. Quoted in Macrae, *John von Neumann*.
11. Von Neumann Whitman, *The Martian's Daughter.*
12. 'It's basically Angry Birds,' says historian Thomas Haigh. http://opentranscripts.org/transcript/working-on-eniac-lost-labors-information-age/.
13. Quoted in Dyson, *Turing's Cathedral*.
14. Quoted in ibid.
15. Quoted in Macrae, *John von Neumann*.
16. Von Neumann Whitman, *The Martian's Daughter*.

17. Quoted in Dyson, *Turing's Cathedral.*
18. https://libertyellisfoundation.org/passenger-details/czoxMzoiOTAxMTk4OTg3MDU0MSI7/czo4OiJtYW5pZmVzdCI7.
19. Details of Meitner's life drawn from Ruth Lewin Sime, 1996, *Lise Meitner: A Life in Physics*, University of California Press, Berkeley.
20. John von Neumann, 2005, *John von Neumann: Selected Letters*, ed. Miklós Rédei, American Mathematical Society, Providence, R.I.
21. Subrahmanyan Chandrasekhar and John von Neumann, 1942, 'The Statistics of the Gravitational Field Arising from a Random Distribution of Stars. I. The Speed of Fluctuations', *Astrophysical Journal*, 95 (1942), pp. 489–531.
22. Thomas Haigh and Mark Priestly have recently made the case that von Neumann was not much influenced by Turing when it came to computer design, based on the text of three lectures they discovered: 'Von Neumann Thought Turing's Universal Machine Was "Simple and Neat". But That Didn't Tell Him How to Design a Computer', *Communications of the ACM*, 63(1) (2020), pp. 26–32.
23. The name of the committee was not an acronym. A member of the committee, John Cockcroft, had received a cryptic telegram from Lise Meitner via an English friend of hers: 'MET NIELS AND MARGRETHE RECENTLY BOTH WELL BUT UNHAPPY ABOUT EVENTS PLEASE INFORM COCKCROFT AND MAUD RAY KENT'. Cockcroft puzzled over the contents of the telegram. With a 'y' replaced with an 'i', he reasoned that the last three words were an anagram for 'radium taken'. Were the Germans stockpiling the radioactive substance for work on a reactor or bomb? The committee was suitably galvanized by Meitner's cunning warning, taking for its name the first word of the mysterious triplet in her message: MAUD. Many years after Cockcroft and others puzzled over the import of her words, it would emerge that Meitner's message was never meant to be a call to action but was addressed to the former governess of Bohr's children. Her name was Maud Ray. She lived in Kent. Meitner, the committed pacifist, had inadvertently helped to launch and energize the pursuit of the atomic bomb in two countries.
24. Two great accounts of the development of the atom bomb are Richard Rhodes, 2012, *The Making of the Atom Bomb*, Simon & Schuster, London, and the more recent Jim Baggott, 2012, *Atomic: The First War of Physics and the Secret History of the Atom Bomb: 1939–49*, Icon Books, London.
25. Kenneth D. Nichols, quoted in Peter Goodchild, 1980, *J. Robert Oppenheimer: Shatterer of Worlds*. Houghton Mifflin, New York.

26. Quoted in Rhodes, *The Making of the Atom Bomb.*
27. The technical problems encountered by the Manhattan Project and some of von Neumann's contribution to solving them are from Lillian Hoddeson, Paul W. Henriksen, Roger A. Meade and Catherine Westfall, 1993, *Critical Assembly: A Technical History of Los Alamos during the Oppenheimer Years, 1943–1945*, Cambridge University Press, Cambridge.
28. The discovery was kept secret until after the war.
29. Quoted in Hoddeson et al., *Critical Assembly.*
30. John von Neumann, 1963, *Oblique Reflection of Shocks*, in *John von Neumann: Collected Works*, ed. A. H, Taub, vol. 6: *Theory of Games, Astrophysics, Hydrodynamics and Meteorology*, Pergamon Press, Oxford.
31. Hoddeson et al., *Critical Assembly.*
32. Kistiakowsky claimed to have taught von Neumann how to play poker at Los Alamos, only stopping when 'he began to play better than I'. Yet von Neumann was familiar enough with the game to mention poker in a paper on game theory he published in 1928 and by the time he met Kistiakowsky, he had finished *Theory of Games and Economic Behavior*, which includes various analyses of the game. Perhaps von Neumann, who was a notoriously poor player, was trying to lull his friend into a false sense of security. https://www.manhattanprojectvoices.org/oral-histories/george-kistiakowskys-interview.
33. In the months that followed, experiments confirmed that reactor-produced samples contained a high-proportion of plutonium-240, which decays faster than the plutonium-239 needed at Los Alamos.
34. The soccer balls of the time did not, however, resemble truncated icosahedrons. Most were made of up to 18 leather panels stitched together.
35. Arjun Makhijani, '"Always" the Target?', *Bulletin of the Atomic Scientists*, 51(3) (1995), pp. 23–7.
36. 'Personal Justice Denied: Report of the Commission on Wartime Relocation and Internment of Civilians', National Archives. Government Printing Office, Washington, D.C., December 1982, https://www.archives.gov/research/japanese-americans/justice-denied.
37. Von Neumann's notes on the Target Committee's agenda are taken from Macrae, *John von Neumann.*
38. http://www.dannen.com/decision/targets.html.
39. https://www.1945project.com/portfolio-item/shigeko-matsumoto/.
40. The Committee for the Compilation of Materials on Damage Caused by the Atomic Bombs in Hiroshima and Nagasaki, 1981, *Hiroshima and Nagasaki: The Physical, Medical, and Social Effects of the Atomic Bombings*, Basic Books, New York.

41. Freeman Dyson, 1979. *Disturbing the Universe*, Harper and Row, New York.
42. Quoted in von Neumann Whitman, *The Martian's Daughter*.
43. Quoted in Dyson, *Turing's Cathedral*.
44. Interview with the author. 14 January 2019.
45. Von Neumann Whitman, *The Martian's Daughter*.
46. Interview with the author.
47. German A. Goncharov, 'Thermonuclear Milestones: (1) The American Effort', *Physics Today*, 49(11) (1996), pp. 45–8.
48. https://www.globalsecurity.org/wmd/intro/classical-super.htm.
49. Goncharov, 'Thermonuclear Milestones'.
50. German A. Goncharov, 'Main Events in the History of the Creation of the Hydrogen Bomb in the USSR and the USA', *Physics – Uspekhi*, 166 (1996), pp. 1095–1104.

CHAPTER 5: THE CONVOLUTED BIRTH OF THE MODERN COMPUTER

1. From her unfinished memoirs, quoted extensively in Dyson, *Turing's Cathedral*. This chapter is particularly indebted to Dyson; Thomas Haigh, Mark Priestley and Crispin, Rope, 2016, *ENIAC in Action: Making and Remaking the Modern Computer*, MIT Press, Cambridge, Mass., and William Aspray, 1990, *John von Neumann and the Origins of Modern Computing*, MIT Press, Cambridge, Mass.
2. A slightly longer excerpt is quoted in Leonard, *Von Neumann, Morgenstern, and the Creation of Game Theory Cambridge University Press, Cambridge*.
3. See Macrae, *John von Neumann*.
4. Earl of Halsbury, 'Ten Years of Computer Development', *Computer Journal*, 1 (1959), pp. 153–9.
5. Brian Randell, 1972, *On Alan Turing and the Origins of Digital Computers*, University of Newcastle upon Tyne Computing Laboratory, Technical report series.
6. Quoted in Aspray, *John von Neumann and the Origins of Modern Computing*.
7. Interview with William Aspray, quoted in ibid.
8. Hermann H. Goldstine, 1972, *The Computer from Pascal to von Neumann*, Princeton University Press, Princeton.

9. Herman Goldstine, interview with Albert Tucker and Frederik Nebeker, 22 March 1985, https://web.math.princeton.edu/oral-history/c14.pdf.
10. Harry Reed, 18 February 1996, ACM History Track Panel, quoted in Thomas J. Bergin (ed.), 2000, *50 Years of Army Computing: From ENIAC to MSRC*, Army Research Lab Aberdeen Proving Ground MD.
11. Their contribution was discovered by the historian Thomas Haigh and his colleagues. See Haigh et al., *ENIAC in Action*.
12. http://opentranscripts.org/transcript/working-on-eniac-lost-labors-information-age/.
13. From dissertation work by Anne Fitzpatrick. Quoted in Haigh et al., *ENIAC in Action*.
14. Quoted in ibid.
15. Ibid.
16. Originally circulated in 1945, a carefully edited version was published by Michael D. Godfrey in 1993. John von Neumann, 'First Draft of a Report on the EDVAC', *IEEE Annals of the History of Computing*, 15 (1993), pp. 27–75.
17. Wolfgang Coy, 2008, *The Princeton Companion to Mathematics*, Princeton University Press, Princeton.
18. The details of Gödel's life and work here are drawn largely from John W. Dawson, 1997, *Logical Dilemmas: The Life and Work of Kurt Gödel*, A. K. Peters, Wellesley, Mass., and Rebecca Goldstein, 2005, *Incompleteness: The Proof and Paradox of Kurt Gödel*. W. W. Norton & Company, New York.
19. In a syllogism, like the one below, a pair of premises leads to a conclusion:

    All men are mortal
    Socrates is a man
    Therefore, Socrates is mortal

    In the symbols of first-order logic, this becomes:

    $\forall x(M(x) \rightarrow P(x))$
    $M(a)$
    $P(a)$

    . . . where

    $\forall$ means 'every', $\rightarrow$ means 'implies' or 'is'
    M(a), which is read as '*M* of *a*', means that any *a* has the property *M*.
    M = the property of being a man
    P = the property of being mortal
    a is Socrates.

20. David Hilbert and Wilhelm Ackermann, 1928, *Grundzüge der theoretischen Logik*, Julius Springer, Berlin (later translated as *Principles of Mathematical Logic*).
21. E.g. set M = the property of being an orange, P = the property of being the colour green, and insert into the syllogism above.
22. Andrew Wiles of the University of Oxford proved Fermat's last theorem in 1994, more than 350 years after Pierre de Fermat stated the problem and asserted he had a 'marvellous proof' which his margin was too narrow to contain. Wiles's proof ran to over 100 pages. Goldbach's conjecture remains unproven.
23. 'Gödel began the same way, assigning a unique number to each symbol in the *Principia*': after that he used prime numbers to denote the position of a particular symbol (the first five primes are 2, 3, 5, 7 and 11). He then raised each prime to the power of the symbol number of the symbol at that position in the statement. Multiplying these together produces the Gödel number of that statement. So, for example, let's define the following symbol numbers: $M = 1$, $a = 2$, $(= 3$, and $) = 4$. Then the Gödel number of the expression $M(a)$ is $2^1 \times 3^3 \times 5^2 \times 7^4 = 3{,}241{,}350$. As you can see, Gödel numbers soon get very large indeed. The so-called 'fundamental theorem of arithmetic' states that every integer greater than 1 is either prime or is a unique product of primes. That means in turn every statement has a unique Gödel number, and by factorizing that number, we can *always* recover the encoded statement.
24. 'The Gödelian strange loop that arises in formal systems in mathematics ... is a loop that allows such a system to "perceive itself", to talk about itself, to become "self-aware", and in a sense it would not be going too far to say that by virtue of having such a loop, a formal system *acquires a self*.' Douglas R. Hofstadter, 1979, *Gödel, Escher, Bach: An Eternal Golden Braid*, Basic Books, New York. Quote from the preface to the twentieth-anniversary edition.
25. Not that he knew or cared much about that.
26. Minutes of the Institute for Advanced Study Electronic Computing Project, Meeting 1, 12 November 1945, IAS, quoted in Dyson, *Turing's Cathedral*.
27. Martin Davis, 2000, *The Universal Computer: The Road from Leibniz to Turing*, W. W. Norton & Company, New York.
28. Kurt Gödel, 'Über formal unentscheidbare Sätze der Principia Mathematica und verwandter Systeme I', *Monatshefte für Mathematik und Physik*, 38 (1931), pp. 173–98.

29. Von Neumann, *Selected Letters*.
30. Any details of Turing's life are drawn from Andrew Hodges, 2012, *Alan Turing: The Enigma. The Centenary Edition*, Princeton University Press, Princeton. My brief description of Turing's paper is abridged from Charles Petzold, 2008, *The Annotated Turing: A Guided Tour Through Alan Turing's Historic Paper on Computability and the Turing Machine*, Wiley, Hoboken, and 'Computable Numbers: A Guide', in Jack B. Copeland (ed.), 2004, *The Essential Turing: Seminal Writings in Computing, Logic, Philosophy, Artificial Intelligence, and Artificial Life plus The Secrets of Enigma*, Oxford University Press, Oxford.
31. Alonzo Church, 'A Note on the *Entscheidungsproblem*', *Journal of Symbolic Logic*, 1(1) (1936), pp. 40–41.
32. Turing, 'On Computable Numbers'.
33. He summarizes the behaviour of the machine in a table that looks something like this:

Starting *m*-configuration	Symbol read	Action	Final *m*-configuration
a	blank	P0, R	b
b	blank	R	c
c	blank	P1, R	d
d	blank	R	a

where P0 means 'print a 0', P1 means 'print a 1' and R means move right one square.

34. Charles Petzold describes Turing machines that can, for example, add and multiply two binary numbers together. See Petzold, *The Annotated Turing*.
35. Supplied with a copy of the standard description, all the universal machine needs to start work is the initial state of the Turing machine it is imitating and the first symbol that machine is supposed to read. It begins by scanning the standard description on its tape for the symbol and initial state that it has been given. Once it has found this configuration, the universal machine can read what it should do next: the coded *m*-description supplies the symbol to be printed, how its head should move and the next state. With this information to hand, the universal machine winds its tape back to where it started, prints the next symbol and the new state of the Turing machine it is simulating. Now the universal machine has a new symbol and state to look for in the standard description on its tape. This cyclical 'search and match',

followed by print symbol and switching to next state, can continue for ever – or until the 'program' on its tape terminates. The output of the universal machine is not identical to that of the machine it is emulating. The universal machine, for instance, requires extra space for printing states. Nonetheless, it *does* reproduce the expected sequence exactly – even if there are gaps in between for 'rough working'.

36. John von Neumann, 1963, 'The General and Logical Theory of Automata', lecture given at the Hixon Symposium on Cerebral Mechanisms in Behaviour, 20 September 1948, reproduced in *Collected Works*, vol. 5: *Design of Computers, Theory of Automata and Numerical Analysis*, Pergamon Press, Oxford.
37. Briefly, Turing's strategy is this: he begins by considering whether it is possible for some Turing machine to tell from another machine's standard description whether it will print digits for ever or halt. A little confusingly, he calls machines that print for ever 'circle free' and those that eventually stop as 'circular'. Turing then proves that no machine exists that is capable of deciding if another, arbitrary machine is circle free. Any such machine, he shows, gets stuck in a loop when trying to assess its own standard description. Next, he likewise demonstrates that no machine can be concocted that can decide whether another machine ever prints a given symbol – '0', for example. He does so by showing that this would also imply the existence of a machine that could determine whether a machine is circle free. Since he has already shown no machine can determine if another prints for ever by analysing its standard description, there can also be no machine that determines if another prints '0'. Finally, Turing constructs a rather complex statement in first-order logic that, in effect, says, '0 appears somewhere on the tape of machine *M*'. He calls this formula Un(*M*), for 'undecidable'. Now, imagine a machine that can determine whether any statement in first-order logic can be proved using first-order logic. The philosopher Jack Copeland calls this machine 'Hilbert's dream'. Alas, when Hilbert's dream attempts to chunter its way through Un(*M*) to decide if it is provable, the wheels come off. For Turing has already demonstrated that *no* machine can ever determine this. Hilbert's dream disappears in a puff of logic.
38. From Klára von Neumann's papers. Quoted in Dyson, *Turing's Cathedral*.
39. Haigh quashes some of the wilder claims made for Turing in Thomas Haigh, 'Actually, Turing Did Not Invent the Computer', *Communications of the ACM*, 57(1) (2014), pp. 36–41.

40. Copeland, *The Essential Turing.*
41. *First Draft of a Report on the EDVAC.* A carefully proofread and edited version was published by Michael D. Godfrey, *IEEE Annals of the History of Computing*, 15(4) (1993), pp. 27–75.
42. W. S. McCulloch and W. Pitts, 1943, 'A Logical Calculus of the Ideas Immanent in Nervous Activity', *Bulletin of Mathematical Biophysics*, 5 (1943), pp. 115–33.
43. Goldstine to von Neumann, 15 May, 1945, quoted in Haigh et al., *ENIAC in Action.*
44. From ibid.
45. The computer's 'main memory' or 'primary storage' loses data as soon as the computer powers down. The auxiliary memory, provided by a hard drive or flash memory, holds data until overwritten.
46. John W. Mauchly, 'Letter to the Editor', *Datamation*, 25(11) (1979), https://sites.google.com/a/opgate.com/eniac/Home/john-mauchly.
47. J. Presper Eckert, oral history interview by Nancy B. Stern, 28 October 1977, http://purl.umn.edu/107275.
48. From an unpublished book, quoted in Haigh et al., *ENIAC in Action.* This remains a controversial statement in computer history. In their thorough, recent work on the ENIAC, Haigh and his colleagues conclude: 'Our best interpretation of the evidence is that, by editing, assembling, and extending ideas discussed at the joint meetings with the ENIAC team, von Neumann established, for the first time, the EDVAC architecture as a unified whole. See ibid.
49. Von Neumann to Aaron Townshend, 6 June 1946, quoted in Aspray, *John von Neumann and the Origins of Modern Computing.*
50. John von Neumann, deposition concerning EDVAC report, n.d. [1947], IAS, quoted in Dyson, *Turing's Cathedral.*
51. Von Neumann to Frankel, 29 October 1946, quoted in Aspray, *John von Neumann and the Origins of Modern Computing.*
52. I. J. Good, 1970, 'Some Future Social Repercussions of Computers', *International Journal of Environmental Studies*, 1 (1970), pp. 67–79.
53. Eckert and Mauchly left the Moore School in March 1946 after refusing to sign new agreements to hand over their inventions to the University of Pennsylvania. They founded their own computer company, producing the UNIVAC I, the first of a series of machines for business and military applications, which was sold at a loss to the US Census Bureau. The Eckert–Mauchly Computer Corporation soon ran into severe financial difficulties, which were exacerbated when Mauchly was falsely accused of having communist sympathies. Some lucrative defence

contracts were cancelled, and the company was put up for sale. The firm was acquired in 1950 by Remington Rand, which after a merger became Sperry Rand. As the new owners of Eckert and Mauchly's patents, Sperry Rand would defend them, unsuccessfully, in court. Despite the devastating conclusion of that battle, Sperry Rand lives on today: after a takeover in 1986 by Burroughs, a business machines maker, the combined company became the computer giant Unisys.

54. Norbert Wiener to John von Neumann, 24 March 1945. Quoted in both Macrae, *John von Neumann*, and Dyson, *Turing's Cathedral*.
55. Minutes of the Regular Meeting of the Board of Trustees, Institute for Advanced Study, 19 October 1945, quoted in Goldstine, *The Computer from Pascal to von Neumann*.
56. Dyson, *Disturbing the Universe*.
57. Klára von Neumann, *Johnny*, quoted in Dyson, *Turing's Cathedral*.
58. Minutes of the Regular Meeting of the Board of Trustees, Institute for Advanced Study, 19 October 1945, quoted in Goldstine, *The Computer from Pascal to von Neumann*.
59. John von Neumann to Lewis Strauss, October 1945, quoted in Andrew Robinson (ed.), 2013, *Exceptional Creativity in Science and Technology: Individuals, Institutions, and Innovations*, Templeton Press, West Conshohocken, Pennsylvania.
60. All details of the ENIAC's conversion and programming are from Haigh et al., *ENIAC in Action*.
61. In conversation with N. Cooper: see N. G. Cooper et al. (eds.), 1989, *From Cardinals to Chaos: Reflection on the Life and Legacy of Stanislaw Ulam*, Cambridge University Press, Cambridge.
62. Quoted in Dyson, *Turing's Cathedral*.
63. Klára von Neumann, *c.*1963, quoted in Dyson, *Turing's Cathedral*.
64. Ulam, quoted in Roger Eckhardt, 1987, *Stan Ulam, John von Neumann, and the Monte Carlo Method*, *Los Alamos Science*, 15, Special Issue (1987), pp. 131–7, https://permalink.lanl.gov/object/tr?what=info:lanl-repo/lareport/LA-UR-88-9068.
65. Ulam was playing Canfield (US) or Demon (UK) solitaire, which is notoriously difficult to win.
66. Ulam testimony during ENIAC patent trial, quoted in Dyson, *Turing's Cathedral*.
67. Haigh et al. were not able to find the program code for the first Monte Carlo run, but code listings for later runs were written in Klári's hand.
68. 'Eniac, the only electronic computer among the four "mathematical brains" now in use, is being converted so that it can handle without

resetting all types of mathematical problems to which it is adapted.' The changes would, the *Times* says, give ENIAC 'a substantial part of the efficiency which is being built into the Edvac' (Will Lissner, '"Brain" Speeded Up for War Problems', *New York Times*, 13 December 1947), and 'Under changes being made, it is hoped to boost Eniac's actual weekly output from the equivalent of 10,000 man-hours to 30,000' (Will Lissner, 'Mechanical "Brain" Has Its Troubles', *New York Times*, 14 December 1947).

69. Nicholas Metropolis, Jack Howlett and Gian-Carlo Rota (eds.), 1980, *A History of Computing in the Twentieth Century*, Academic Press, New York.
70. John von Neumann, 1951, 'Various Techniques Used in Connection with Random Digits', https://mcnp.lanl.gov/pdf_files/nbs_vonneumann.pdf.
71. S. Ulam to von Neumann, 12 May 1948, Putnam, New York quoted in Dyson, *Turing's Cathedral.*
72. Von Neumann to S. Ulam, 11 May 1948, quoted in Haigh et al., *ENIAC in Action.*
73. Klári von Neumann to S. Ulam, 12 June 1948, quoted in ibid.
74. See Haigh et al., *ENIAC in Action.*
75. Von Neumann to S. Ulam, 18 November 1948, quoted in ibid.
76. Klári von Neumann to the Ulams, quoted in Dyson, *Turing's Cathedral.*
77. The full name of MANIAC I was Mathematical Analyzer Numerical Integrator and Computer Model I.
78. See Haigh et al., *ENIAC in Action.*
79. Julian Bigelow interview with Nancy Stern, quoted in Dyson, *Turing's Cathedral.*
80. Dyson, *Turing's Cathedral.*
81. Interview with Richard R. Mertz, quoted in ibid.
82. Interview with Nancy Stern, quoted in ibid.
83. Stanley A. Blumberg and Gwinn Owens, 1976, *Energy and Conflict: The Life and Times of Edward Teller,*.
84. John von Neumann, 'Defense in Atomic War', *Scientific Bases of Weapons, Journal of American Ordnance Association*, 6(38) (1955), pp. 21–3, reprinted in *Collected Works*, vol. 6.
85. Julian Bigelow, 'Computer Development at the Institute for Advanced Study', in Metropolis et al. (eds.), *A History of Computing in the Twentieth Century*, pp. 291–310.

CHAPTER 6: A THEORY OF GAMES

1. Von Neumann Whitman, *The Martian's Daughter.*
2. Ibid.
3. Ibid.
4. Jacob Bronowski, *The Ascent of Man*, Little, Brown, 1975.
5. This chapter is indebted to Robert Leonard, *Von Neumann, Morgenstern and the Creation of Game Theory: From Chess to Social Science, 1900–1960*, Cambridge University Press, Cambridge and William Poundstone, *Prisoner's Dilemma: John von Neumann, Game Theory and the Puzzle of the Bomb,* Doubleday, New York. For a brief non-mathematical introduction to game theory, try Ken Binmore, *Game Theory: A Very Short Introduction*, Oxford University Press, Oxford.
6. 'I was in one of those moods where danger is attractive,' Lasker wrote after a match against Dutch champion Abraham Speijer. 'Hence I plunged from the start into a combination the outcome of which was exceedingly doubtful.' Needless to say, Lasker came out on top.
7. Quoted in Leonard, *Von Neumann, Morgenstern and the Creation of Game Theory.*
8. Emanuel Lasker, [1906/7], *Kampf*, Lasker's Publishing Co., New York, reprinted in 2001 by Berlin-Brandenburg, Potsdam, quoted in ibid.
9. Emanuel Lasker, *Lasker's Manual of Chess*, New York: Dover, 1976 (original: *Lehrbuch des Schachspiels*, 1926; first English translation, 1927)], quoted in ibid.
10. John von Neumann, 'Zur Theorie der Gesellschaftsspiele', *Mathematische Annalen*, 100 (1928), pp. 295–320. Translation by Sonya Bargmann, 'On the Theory of Games of Strategy', *Contributions to the Theory of Games*, 4 (1959), pp. 13–42.
11. Von Neumann Whitman, *The Martian's Daughter*.
12. Von Neumann, 'Zur Theorie der Gesellschaftsspiele'.
13. That no real player is actually perfectly rational does not matter here. Perfectly straight lines do not exist either, but geometry has still proved quite useful in the real world.
14. On von Neumann's 1928 proof and the priority dispute with Borel see Tinne Hoff Kjeldsen, 'John von Neumann's Conception of the Minimax Theorem: A Journey Through Different Mathematical Contexts', *Archive for History of Exact Sciences*, 56 (2001), pp. 39–6.
15. Maurice Fréchet, 'Emile Borel, Initiator of the Theory of Psychological Games and Its Application', *Econometrica*, 21 (1953), pp. 95–6.

16. Maurice Fréchet, 'Commentary on the Three Notes of Emile Borel', *Econometrica*, 21 (1953), pp. 118–24.
17. John von Neumann, 'Communication on the Borel Notes', *Econometrica* 21 (1953), pp. 124–5.
18. Halperin Interview, The Princeton Mathematics Community in the 1930s, Transcript Number 18 (PMC18), quoted in Leonard, *Von Neumann, Morgenstern, and the Creation of Game Theory.*
19. Péter Rózsa, *Játék a Végtelennel*, 1945. Translated by Z. P. Dienes, *Playing with Infinity: Mathematical Explorations and Excursions*, Dover Publications, New York.
20. John von Neumann, 'Über ein ökonomisches Gleichungssystem und eine Verallgemeinerung des Brouwerschen Fixpunktsatzes', *Ergebnisse eines Mathematische Kolloquiums*, 8 (1937), ed. Karl Menger, pp. 73–83,
21. Translated as 'A Model of General Economic Equilibrium', *Review of Economic Studies*, 13 (1945), pp. 1–9.
22. Macrae, *John von Neumann.*
23. The function $f(x) = x$ is all fixed points, for example, while $f(x) = 7$ has one at $x = 7$.
24. John von Neumann, 'The Impact of Recent Development in Science on the Economy and Economics', speech delivered, published in *Looking Ahead*, 4 (1956), also in A. Bródy and T. Vámos (eds.), 1995, *The Neumann Compendium*, World Scientific, London.
25. John von Neumann to Oskar Morgenstern, 8 October 1947, quoted in Oskar Morgenstern, 'The Collaboration Between Oskar Morgenstern and John von Neumann on the Theory of Games', *Journal of Economic Literature,* 14(3) (1976), pp. 805–16.
26. See Macrae, *John von Neumann.*
27. E. Roy Weintraub, 'On the Existence of a Competitive Equilibrium: 1930–1954', *Journal of Economic Literature*, 21(1) (1983), pp. 1–39.
28. Sylvia Nasar, 1998, *A Beautiful Mind*, Simon & Schuster, New York.
29. Quoted in Leonard, *Von Neumann, Morgenstern, and the Creation of Game Theory.*
30. Morgenstern diary entries quoted in Leonard, *Von Neumann, Morgenstern, and the Creation of Game Theory.*
31. Oskar Morgenstern, 1928, *Wirtschaftprognose: Eine Untersuchung ihrer Voraussetzungen und Möglichkeiten*, Julius Springer, Vienna. Translation quoted in Leonard, *Von Neumann, Morgenstern, and the Creation of Game Theory*.

32. Morgenstern, 'The Collaboration'.
33. Oskar Morgenstern, *Diary*, 18 November 1938, quoted in Leonard, *Von Neumann, Morgenstern, and the Creation of Game Theory.*
34. Morgenstern, *Diary*, 15 February 1939, quoted in Leonard, *Von Neumann, Morgenstern, and the Creation of Game Theory.*
35. Einstein to Queen Elizabeth of Belgium, 20 November 1933, quoted in Jagdish Mehra, 1975, *The Solvay Conferences on Physics: Aspects of the Development of Physics since 1911*, D. Reidel, Dordrecht.
36. Morgenstern, 'The Collaboration'.
37. Morgenstern, *Diary*, 26 October 1940.
38. Ibid., 22 January 1941.
39. Ibid.
40. Israel Halperin, 1990, 'The Extraordinary Inspiration of John von Neumann', in *Proceedings of Symposia in Pure Mathematics*, vol. 50: *The Legacy of John von Neumann*, ed. James Glimm, John Impagliazzo and Isadore Singer, American Mathematical Society, Providence, R.I., pp. 15–17.
41. Morgenstern, *Diary*, 12 July 1941.
42. Klára von Neumann, unpublished papers, quoted in Leonard, *Von Neumann, Morgenstern, and the Creation of Game Theory*.
43. Morgenstern, *Diary*, 7 August 1941.
44. Leonard, *Von Neumann, Morgenstern, and the Creation of Game Theory*.
45. Morgenstern, *Diary*, 14 April 1942.
46. Morgenstern, 'The Collaboration'.
47. This illustration is my version of the explanation given in Binmore, *Game Theory*.
48. The scale need not go from 0 to 100. A utility score on one scale can be converted to a score on the other in the same way that a temperature measured in degrees Celsius can be converted into degrees Fahrenheit.
49. John von Neumann and Oskar Morgenstern, 1944, *Theory of Games and Economic Behavior.* Princeton University Press, Princeton.
50. Daniel Kahneman, 2011, *Thinking, Fast and Slow*, Farrar, Straus and Giroux, New York.
51. German logician Ernst Zermelo proved in 1912 that from a winning position either black or white can force a win. Unlike von Neumann, he allowed for games with infinitely many moves by ignoring the standard stopping rules of chess and did not use the method of backward induction for his proof. See U. Schwalbe and P. Walker, 'Zermelo

and the Early History of Game Theory', *Games and Economic Behavior*, 34 (2001), pp. 123–37.

52. In this scenario if Holmes goes to Dover, Moriarty's average payout is 60 per cent of 100 (60) if he goes to Dover plus 40 per cent of -50 (-20) if stopped at Canterbury. If Holmes disembarks early, on the other hand, then Moriarty's strategy yields 40 per cent of 100. So in both cases Moriarty's strategy yields 40 utils.
53. Binmore, *Game Theory*.
54. There are 33 'cards' higher than hers and 66 of equal or lower value.
55. The threshold is given by $\frac{H-L}{H} \times 99$.
56. The optimal frequency of bluff bids is $\frac{L}{H+L}$.
57. Binmore, *Game Theory.*
58. I.e. Players A, B and C can form the coalitions (A,B), (B,C) and (A,C).
59. Imagine two players form a coalition and divide the payoff equally, each getting ½. The third player, who now faces receiving a payout of -1, quickly offers one of the others ¾ utils if they team up with him. He would then pocket ¼ utils. But then there would be nothing to stop the newly jilted player (now facing a payout of -1) from offering him more than ¼ utils to form another alliance and so on . . .
60. W Barnaby, 'Do Nations Go to War Over Water?', *Nature*, 458 (2009), pp. 282–3.
61. Von Neumann calls these 'imputations', a term still used in game theory today.
62. Michael Bacharach, 1989, 'Zero-sum Games', in John Eatwell, Murray Milgate and Peter Newman (eds.), *Game Theory*, The New Palgrave, Palgrave Macmillan, London, pp. 253–7.
63. Von Neumann knew Walras's work – his own model of General Economic Equilibrium was an attempt to rectify some of the flaws he saw there.
64. John McDonald, 1950, *Strategy in Poker, Business and War*, W. W. Norton, New York.
65. Ibid.
66. Jacob Marschak, 'Von Neumann and Morgenstern's New Approach to Static Economics', *Journal of Political Economy*, 54 (1946), pp. 97–115.
67. Robert J. Leonard, 'Reading Cournot, Reading Nash: The Creation and Stabilisation of the Nash Equilibrium', *Economic Journal*, 104(424) (1994), pp. 492–511.

68. Ibid.
69. William F. Lucas, 'A Game with No Solution', *Bulletin of the American Mathematics Society*, 74 (1968), pp. 237–9.
70. Gerald L. Thompson, 1989, 'John von Neumann', in Eatwell et al. (eds.), *Game Theory*, pp. 242–52.
71. The story of the controversy over the award is told in Nasar, *A Beautiful Mind*.
72. See John McMillan, 'Selling Spectrum Rights', *Journal of Economic Perspectives*, 8(3) (1994), pp. 145–62. Early competitions to assign public assets to companies were known as 'beauty contests'. Each firm submitted a lengthy document explaining why their bid was the best, and a panel of officials decided the winner. Unfortunately, the officials charged with the task could have no real idea what the asset was really worth to the competing businesses – and the businesses had no incentive to tell them. That method was replaced with lotteries, which were also gamed. One winner in 1989 sold their licence to operate mobile phones on Cape Cod to Southwestern Bell for $41 million.
73. Paul Milgrom, 'Putting Auction Theory to Work: The Simultaneous Ascending Auction', *Journal of Political Economy*, 108(2) (2000), pp. 245–72. In a simultaneous ascending auction, bidding progresses in rounds, with the current standings revealed between rounds. The auction is over when there are no new bids on any licence. There are penalties for withdrawing bids to help keep bidders honest. Auctioning off licences in batches rather than sequentially prevents a firm from driving up prices of licences sold earlier so that competitors cannot afford to bid for those sold later. The setup is good for bidders too – they can adjust their bids to collect bundles of licences that complement each other.
74. See e.g. Thomas Hazlett, 2009, 'U.S. Wireless License Auctions: 1994–2009', https://www.accc.gov.au/system/files/Hazlett%2C%20Thomas%20%28Auctions%20Paper%29.pdf, and a complete list of FCC auctions here: https://capcp.la.psu.edu/data-and-software/fcc-spectrum-auction-data.
75. Ostrom was the only woman to win the prize until Esther Duflo in 2019.
76. Derek Wall, 2014, *The Sustainable Economics of Elinor Ostrom: Commons, Contestation and Craft*, Routledge, London.
77. Elinor Ostrom, 'Design Principles of Robust Property Rights Institutions: What Have We Learned?', in Gregory K. Ingram and Yu-Hung Hong (eds.), 2009, *Property Rights and Land Policies*, Lincoln Institute of Land Policy, Columbia University Press, New York.

78. See Kahneman, *Thinking, Fast and Slow*.
79. I am indebted to Michael Ostrovsky for his help with present-day applications of game theory and his magisterial round-up of thirty years of work in the field.
80. For any particular search term, advertisers submit bids stating the maximum they are willing to pay each time someone clicks through to their web page. The highest bidder has their ad shown most prominently on the search results page – but pays the second-highest bid. The second-highest bidder secures the second-most prominent position on the results page – and pays the third-highest bid, and so on. Designing and perfecting these 'generalized second price' auctions brought the first wave of game theorists into tech firms. For a brief history and review of such auctions, see one of the most cited economics papers of this century: Benjamin Edelman, Michael Ostrovsky and Michael Schwarz 'Internet Advertising and the Generalized Second-Price Auction: Selling Billions of Dollars Worth of Keywords', *American Economic Review*, 97(1) (2007), pp. 242–59.
81. W. D. Hamilton, 1996, *Narrow Roads of Gene Land*, vol. 1: *Evolution of Social Behaviour*, Oxford University Press, Oxford.
82. For the remarkable story of Price's quest to understand altruism, see Oren Harman, 2010, *The Price of Altruism: George Price and the Search for the Origins of Kindness*, Bodley Head, London.
83. Elinor Ostrom, 2012, 'Coevolving Relationships between Political Science and Economics', *Rationality, Markets and Morals*, 3 (2012), pp. 51–65.

CHAPTER 7: THE THINK TANK BY THE SEA

1. Details of the history of the RAND Corporation are mainly from Fred Kaplan, 1983, *The Wizards of Armageddon*, Stanford University Press, Stanford, and David Jardini, 2013, *Thinking Through the Cold War: RAND, National Security and Domestic Policy, 1945–1975*, Smashwords, as well as Poundstone, *Prisoner's Dilemma* and Alex Abella, 2008, *Soldiers of Reason: The RAND Corporation and the Rise of the American Empire,* Harcourt, San Diego, Calif. Daniel Bessner, 2018, *Democracy in Exile: Hans Speier and the Rise of the Defense Intellectual*, Cornell University Press, Ithaca, offers another perspective on RAND. The definitive intellectual history of game theory is Paul Erickson, 2015, *The World the Game Theorists Made*, The University of Chicago Press, Chicago.

2. 'The RAND Hymn', words and music by Malvina Reynolds, copyright 1961 Schroder Music Company, renewed 1989.
3. Kaplan, *The Wizards of Armageddon.*
4. Quoted in Jacob Neufeld, 1990, *The Development of Ballistic Missiles in the United States Air Force, 1945–1960*, United States Government Printing Office, Washington, D.C.
5. Kaplan, *Wizards of Armageddon.*
6. Ibid.
7. Ibid.
8. H. H. Arnold, 1949, *Global Mission*, Harper & Brothers, New York.
9. Full report available: https://www.governmentattic.org/TwardNewHorizons.html See also https://apps.dtic.mil/dtic/tr/fulltext/u2/a954527.pdf.
10. Interview with Collbohm by Martin Collins and Joseph Tatarewicz, 28 July 1987, https://www.si.edu/media/NASM/NASM-NASM_AudioIt-0000066400DOCS.pdf.
11. Quoted in Abella, *Soldiers of Reason.*
12. Leonard, *Von Neumann, Morgenstern, and the Creation of Game Theory.*
13. Quoted in Larry Owens, 1989, *Mathematicians at War: Warren Weaver and the Applied Mathematics Panel, 1942–45*, ed. David Rower and John McCleary, Academic Press, Boston.
14. Bernard Lovell, 1988, 'Blackett in War and Peace', *The Journal of the Operational Research Society*, 39(3) (1988), pp. 221–33.
15. All quotations here are from Erickson, *The World the Game Theorists Made.*
16. Quoted in Abella, *Soldiers of Reason.*
17. Kaplan, *The Wizards of Armageddon.*
18. John Williams to von Neumann 16 December 1947, quoted in Erickson, *The World the Game Theorists Made.*
19. Poundstone, *Prisoner's Dilemma.*
20. Ibid.
21. See George B. Dantzig, 'The Diet Problem', *Interfaces*, 20(4) (1990), pp. 43–7. Dantzig later tried using the algorithm to lose weight, promising his wife that he would follow whatever optimal diet RAND's computer suggested. Unfortunately, owing to an oversight during optimization of the program, the computer recommended that Dantzig drink 500 gallons of vinegar.
22. https://apps.dtic.mil/dtic/tr/fulltext/u2/a157659.pdf. George B. Dantzig, 1985, *Impact of Linear Programming on Computer Development*, Department of Operations Research, Stanford University.

23. Interview with Robert Leonard, 27 February 1990, quoted in Leonard, *Von Neumann, Morgenstern, and the Creation of Game Theory*.
24. Willis H. Ware, 2008, *RAND and the Information Evolution: A History in Essays and Vignettes*, RAND Corporation.
25. Clay Blair Jr, 'Passing of a Great Mind', *Fortune*, 25 February 1957, p. 89.
26. Quoted in Leonard, *Von Neumann, Morgenstern, and the Creation of Game Theory*.
27. Much of this work is collected and summarized in Melvin Dresher, 1961, *Games of Strategy: Theory and Applications*, available as RAND Corporation document number CB-149-1 (2007).
28. This episode is related in Leonard, *Von Neumann, Morgenstern, and the Creation of Game Theory*.
29. Hans Speier interview with Martin Collins, 5 April 1988, https://www.si.edu/media/NASM/NASM-NASM_AudioIt-0000318 1DOCS.pdf.
30. Mathematician Joseph Malkevitch gives an excellent example of how Shapley values work by considering the case of three (fictional) towns ordered to clean up their sewage outflow. See http://www.ams.org/publicoutreach/feature-column/fc-2016-09.
31. Quoted in Alvin E. Roth, 'Lloyd Shapley (1923–2016)', *Nature*, 532 (2016), p. 178.
32. D. Gale and L. S. Shapley, 'College Admissions and the Stability of Marriage', *American Mathematical Monthly*, 69 (1962), pp. 9–15.
33. Sylvia Nasar, 1998, *A Beautiful Mind*, Simon & Schuster, New York.
34. Ibid.
35. J. F. Nash, 'The Bargaining Problem', *Econometrica*, 28 (1950), pp. 155–62.
36. See Leonard, *Reading* Cournot, *Reading* Nash: 'Above all, as Shubik (1991) reports, von Neumann "hated it!", clearly finding it foreign to his whole conception of game theory.'
37. Email from John Nash to Robert Leonard, 20 February 1993, quoted in Nasar, *A Beautiful Mind*.
38. See, for example, Herman Goldstine interview with Albert Tucker and Frederik Nebeker, 22 March 1985, https://web.math.princeton.edu/oral-history/c14.pdf. 'I don't think Johnny suffered brilliant people easily,' says Goldstine. Tucker cites the example of graduate student Harold Kuhn pointing out a possible problem with von Neumann's expanding economy paper. 'Johnny actually got angry,' says Tucker, 'and I really think that he was thinking very fast to get himself out of it.'

39. John F. Nash Jr, 'Equilibrium Points in N-Person Games', *Proceedings of the National Academy of Sciences of the United States of America*, 36 (1) (1950), pp. 48–9.
40. Nasar, *A Beautiful Mind.*
41. Sylvia Nasar claims Nash was not gay: https://www.theguardian.com/books/2002/mar/26/biography.highereducation.
42. Leonard, *Von Neumann, Morgenstern, and the Creation of Game Theory.*
43. Steve J. Heims, 1982, *John von Neumann and Norbert Wiener: From Mathematics to the Technologies of Life and Death*, MIT Press, Cambridge, Mass.
44. Quoted in Dyson, *Turing's Cathedral.*
45. Quoted in ibid.
46. See Leonard, 'Reading Cournot, Reading Nash'.
47. Merrill M. Flood, 1952, *Some Experimental Games*, RAND Research Memorandum RM-789-1.
48. From Poundstone, *Prisoner's Dilemma.*
49. Flood, *Some Experimental Games.*
50. Ibid.
51. This is the method of 'backward induction', also used earlier by von Neumann in *Theory of Games.*
52. Poundstone, *Prisoner's Dilemma.*
53. Evidence of cooperation in the one-shot Prisoner's Dilemma has been found in many studies, e.g. R. Cooper, D. V. DeJong, R. Forsythe and T. W. Ross, 'Cooperation Without Reputation: Experimental Evidence from Prisoner's Dilemma Games', *Games and Economic Behavior*, 12(2) (1996), pp. 187–218, and J. Andreoni and J. H. Miller (1993). 'Rational Cooperation in the Finitely Repeated Prisoner's Dilemma, Experimental Evidence', *The Economic Journal*, 103(418), pp. 570–85.
54. Quoted in von Neumann Whitman, *The Martian's Daughter.*
55. Clay Blair Jr, 'Passing of a Great Mind', *Life Magazine*, 25 February 1957.
56. E.g. Alexander Field, 'Schelling, von Neumann, and the Event That Didn't Occur', *Games*, 5(1) (2014), pp. 53–89.
57. Ulam, 'John von Neumann 1903–1957'.
58. Eugene Wigner, 1957. 'John von Neumann (1903–1957)', *Yearbook of the American Philosophical Society*, later in 1967, *Symmetries and Reflections: Scientific Essays of Eugene P. Wigner*, Indiana University Press, Bloomington.

59. See e.g. R. Buhite and W. Hamel, 'War for Peace: The Question of an American Preventive War against the Soviet Union, 1945–1955', *Diplomatic History*, 14(3) (1990), pp. 367–84.
60. William L. Laurence, 'How Soon Will Russia Have the A-Bomb?', *Saturday Evening Post*, 6 November 1948, p. 182.
61. Buhite and Hamel 'War for Peace'.
62. Quoted in Poundstone, *Prisoner's Dilemma*.
63. Russell spent much of the fifties denying he had ever advocated bombing the Soviets if they did not give up their nuclear ambitions. He became the first president of the Campaign for Nuclear Disarmament in 1958 and in 1961, at age eighty-nine, he was jailed for seven days for organizing an anti-bomb protests in London. Russell finally came clean during an interview with the BBC in 1959. Asked if it was true that he had advocated a preventive war against Soviet Russia, he answered, 'It's entirely true, and I don't repent of it. It was not inconsistent with what I think now. What I thought all along was that a nuclear war in which both sides had nuclear weapons would be an utter and absolute disaster.'
64. John von Neumann to Klára von Neumann, 8 September, 1954, quoted in Dyson, *Turing's Cathedral*.
65. Bernard Brodie, 1959, *Strategy in the Missile Age*, available as RAND Corporation document number CB-137-1 (2007).
66. https://www.manhattanprojectvoices.org/oral-histories/george-kistiakowskys-interview.
67. Oppenheimer was indubitably a supporter of the Communist Party. See Ray Monk, 2012, *Robert Oppenheimer: A Life Inside the Center*, Doubleday, New York and Toronto.
68. Quoted in Macrae, *John von Neumann*. A charitable view of contemporary Britain, which did not make Alan Turing an earl. He was charged with gross indecency and his security clearance was revoked.
69. J. Robert Oppenheimer Personnel Hearings Transcripts, volume XII, https://www.osti.gov/includes/opennet/includes/Oppenheimer%20hearings/Vol%20XII%20Oppenheimer.pdf.
70. John Earl Haynes, Harvey Klehr and Alexander Vassiliev, 2009, 'Enormous: The KGB Attack on the Anglo-American Atomic Project', in *Spies: The Rise and Fall of the KGB in America*, translations by Philip Redko and Steven Shabad, Yale University Press, New Haven.
71. Quoted in Dyson, *Turing's Cathedral*.
72. Erickson, *The World the Game Theorists Made*.

73. Roberta Wohlstetter, 1962, *Pearl Harbor: Warning and Decision*, Stanford University Press, Stanford.
74. Albert Wohlstetter, Fred Hoffman, R. J. Lutz and Henry S. Rowen, 1954, *Selection and Use of Strategic Air Bases*, RAND Corporation, Santa Monica.
75. Albert Wohlstetter, 'The Delicate Balance of Terror', *Foreign Affairs*, 37 (January 1959); an earlier and more complete version dated December 1958 is available as RAND Paper P-1472, https://www.rand.org/pubs/papers/P1472.html.
76. Kaplan, *The Wizards of Armageddon.*
77. The coinage is Alex Abella's.
78. Kaplan, *The Wizards of Armageddon.*
79. Ibid.
80. Herman Kahn and Irwin Mann, 1957, *Game Theory*, https://www.rand.org/pubs/papers/P1166.html.
81. Sharon Ghamari-Tabrizi, 2005, *The Worlds of Herman Kahn: The Intuitive Science of Thermonuclear War*, Harvard University Press, Cambridge, Mass.
82. Herman Kahn, 1960, *On Thermonuclear War*, Princeton University Press, Princeton.
83. The director borrowed so much from *On Thermonuclear War* that Kahn demanded royalties, to which Kubrick replied, 'That's not how it works, Herman!'
84. *Dr Strangelove, or: How I Stopped Worrying and Learned to Love the Bomb*, directed by Stanley Kubrick (1964).
85. James R. Newman, 'Two Discussions of Thermonuclear War', *Scientific American*, March 1961.
86. Kahn joked he gained ten pounds to disprove Newman's doubts about his existence.
87. Thomas C. Schelling, 1958, 'The Strategy of Conflict: Prospectus for a Reorientation of Game Theory', *Journal of Conflict Resolution*, 2(3), pp. 203–64.
88. Thomas C. Schelling, 1960, *The Strategy of Conflict*, Harvard University Press, Cambridge, Mass.
89. Thomas C. Schelling, 'Bargaining, Communication, and Limited War', *Conflict Resolution*, 1(1) (1957), pp. 20, 34.
90. Von Neumann, 'Defense in Atomic War'.
91. Von Neumann was probably referring to one of the recent hydrogen bomb tests, rather than the atom bombs dropped on Hiroshima and Nagasaki, which were a thousand times less powerful.

92. For a summary of US thinking on deterrence see Marc Trachtenberg, 'Strategic Thought in America, 1952–1966', *Political Science Quarterly*, 104(2) (1989), pp. 301–34.
93. https://fas.org/irp/doddir/dod/jp3_72.pdf.

CHAPTER 8: THE RISE OF THE REPLICATORS

1. https://www.youtube.com/watch?v=3KJbrboP8jQ&feature=emb_title.
2. Alex Ellery, 'Are Self-Replicating Machines Feasible?', *Journal of Spacecraft and Rockets*, 53(2) (2016), pp. 317–27.
3. Correspondence with author.
4. What could possibly go wrong?
5. Von Neumann, 'The General and Logical Theory of Automata'.
6. Ibid.
7. John von Neumann, 'The General and Logical Theory of Automata', originally published in Lloyd A. Jeffress (ed.), 1951, *Cerebral Mechanisms in Behavior: The Hixon Symposium*, Wiley, New York.
8. Robert A. Freitas Jr and Ralph C. Merkle, 2004, *Kinematic Self-Replicating Machines*, Landes Bioscience, Georgetown, Texas, http://www.MolecularAssembler.com/KSRM.htm.
9. Von Neumann, 'The General and Logical Theory of Automata'.
10. Erwin Schrödinger, 1944, *What Is Life?*, Cambridge University Press, Cambridge.
11. For a critical appreciation of Schrödinger's *What Is Life?*, see Philip Ball, 2018, 'Schrödinger's Cat among Biology's Pigeons: 75 Years of *What Is Life?*', *Nature*, 560 (2018), pp. 548–50.
12. Sydney Brenner, 1984, 'John von Neumann and the History of DNA and Self-replication', https://www.webofstories.com/play/sydney.brenner/45.
13. Dyson, *Disturbing the Universe.*
14. Arthur W. Burks, 1966, *Theory of Self-reproducing Automata*, University of Illinois Press, Urbana.
15. John G. Kemeny, 'Man Viewed as a Machine', *Scientific American*, 192(4) (1955), pp. 58–67.
16. Philip K. Dick, 'Autofac', *Galaxy*, November 1955.
17. See Lawrence Sutin, [1989], *Divine Invasions: A Life of Philip K. Dick*, Harmony Books, New York.
18. Von Neumann called the struts 'rigid members', an expression that I avoid for reasons that I hope are obvious.

19. He chooses a transmission state with its output on the southern side.
20. Umberto Pesavento, 1995, 'An Implementation of Von Neumann's Self-reproducing Machine', *Artificial Life*, 2(4) (1995), pp. 337–54.
21. For a look at the history of that field, see Steven Levy, 1993, *Artificial Life: A Report from the Frontier Where Computers Meet Biology*, Vintage, New York.
22. Arthur W. Burks, 1966, *Theory of Self-reproducing Automata*, University of Illinois Press, Urbana.
23. Details are from Conway's biography: Siobhan Roberts, 2015, *Genius at Play: The Curious Mind of John Horton Conway*, Bloomsbury, London.
24. Ibid.
25. Ibid.
26. Ibid.
27. Martin Gardner, 2013, *Undiluted Hocus-Pocus: The Autobiography of Martin Gardner*, Princeton University Press, Princeton.
28. Quoted in Levy, *Artificial Life*.
29. These are the AND, OR and NOT operations of Boolean algebra.
30. In 2001, Paul Rendell implemented a Turing machine in Life and, later, a universal version: http://www.rendell-attic.org/gol/tm.htm
31. Quoted in Fred Hapgood, 1987, 'Let There Be Life', *Omni*, 9(7) (1987), http://www.housevampyr.com/training/library/books/omni/OMNI_1987_04.pdf.
32. E. O. Wilson, 1975, *Sociobiology: The New Synthesis*, Harvard University Press, Cambridge, Mass.
33. Levy, *Artificial Life*.
34. Quoted in ibid.
35. Published as Tommaso Toffoli, 1977, 'Computation and Construction Universality of Reversible Cellular Automata', *Journal of Computer and System Sciences*, 15(2), pp. 213–31.
36. Quoted in Levy, *Artificial Life*.
37. Ibid.
38. Fredkin made the case for this in Edward Fredkin, 1990, 'Digital Mechanics: An Informational Process based on Reversible Universal Cellular Automata', *Physica D*, 45 (1990), pp. 254–70.
39. Steven Levy, 'Stephen Wolfram Invites You to Solve Physics', *Wired* (2020), https://www.wired.com/story/stephen-wolfram-invites-you-to-solve-physics/.
40. On the priority dispute between Fredkin and Wolfram, see Levy, *Artificial Life*, and Keay Davidson, 'Cosmic Computer – New Philosophy to Explain the Universe', *San Francisco Chronicle*, 1 July 2002, https://

www.stephenwolfram.com/media/cosmic-computer-new-philosophy-explain-universe/.

41. Steven Wolfram, 2002, *A New Kind of Science*, Wolfram Media, Champagne, Ill.
42. https://www.wolframscience.com/reference/notes/876b.
43. See Wolfram, *A New Kind of Science*.
44. Matthew Cook, 'Universality in Elementary Cellular Automata', *Complex Systems*, 15 (2004), pp. 1–40.
45. Steven Wolfram, 1984, 'Universality and Complexity in Cellular Automata', *Physica D*, 10(1–2), pp. 1–35.
46. Steven Wolfram, 2002, *A New Kind of Science*, Wolfram Media, Champagne, Ill. https://www.wolframscience.com/nks/.
47. Steven Levy, 2002, 'The Man Who Cracked the Code to Everything . . .', *Wired*, 1 June 2002, https://www.wired.com/2002/06/wolfram/.
48. Steven Levy, 'Great Minds, Great Ideas', *Newsweek*, 27 May 2002, p. 59, https://www.newsweek.com/great-minds-great-ideas-145749.
49. https://writings.stephenwolfram.com/2020/04/finally-we-may-have-a-path-to-the-fundamental-theory-of-physics-and-its-beautiful/.
50. See Wolfram's Registry of Notable Universes, https://www.wolframphysics.org/universes/.
51. Adam Becker, 'Physicists Criticize Stephen Wolfram's "Theory of Everything"', *Scientific American*, https://www.scientificamerican.com/article/physicists-criticize-stephen-wolframs-theory-of-everything/.
52. Franz L. Alt, 1972, 'Archaeology of Computers Reminiscences, 1945–1947', *Communications of the ACM*, 15(7) (1972), pp. 693–4, doi:https://doi.org/10.1145/361454.361528.
53. For more on Barricelli see Dyson, *Turing's Cathedral*, chapter 12; Robert Hackett, 'Meet the Father of Digital Life', *Nautilus*, 12 June 2014, https://nautil.us/issue/14/mutation/meet-the-father-of-digital-life; and Alexander, R. Galloway, *Creative Evolution*, http://cultureandcommunication.org/galloway/pdf/Galloway-Creative_Evolution-Cabinet_Magazine.pdf.
54. The experiments using the IAS computer are described in Nils Aall Barricelli, 'Numerical Testing of Evolution Theories. Part I: Theoretical Introduction and Basic Tests', *Acta Biotheoretica*, 16 (1963), pp. 69–98.
55. Ibid.
56. Nils Aall Barricelli to John von Neumann, 22 October 1953, member file on Barricelli, IAS School of Mathematics, members, Ba–Bi, 1933–1977, IAS Archives.

57. Jixing Xia et al., 'Whitefly Hijacks a Plant Detoxification Gene That Neutralizes Plant Toxins', *Cell*, 25 March 2021, https://doi.org/10.1016/j.cell.2021.02.014.
58. Quoted in Levy, *Artificial Life*.
59. Christopher G. Langton, 'Self-reproduction in Cellular Automata', *Physica 10D* (1984), pp. 135–44.
60. Quoted in Levy, *Artificial Life*.
61. Quoted in ibid.
62. Christopher G. Langton, 1990, 'Computation at the Edge of Chaos: Phase Transitions and Emergent Computation', *Physica D*, 42 (1990), pp. 12–37.
63. Christopher G. Langton (ed.), 1989, *Artificial Life*, Santa Fe Institute Studies in the Sciences of Complexity, vol. 6, Addison-Wesley, Reading, Mass.
64. D. G. Gibson et al., 'Creation of a Bacterial Cell Controlled by a Chemically Synthesized Genome', *Science*, 329 (2010), pp. 52–6.
65. See, for example, Nicholas Wade, 'Researchers Say They Created a "Synthetic Cell"', *New York Times*, 20 May 2010.
66. Clyde A. Hutchison III et al., 'Design and Synthesis of a Minimal Bacterial Genome', *Science*, 351 (2016), aad625.
67. Marian Breuer et al., 2019, 'Essential Metabolism for a Minimal Cell', *eLife*, 8 (2019), doi:10.7554/eLife.36842.
68. Kendall Powell, 2018, 'How Biologists Are Creating Life-like Cells from Scratch', *Nature*, 563 (2018), pp. 172–5.
69. Eric Drexler, 1986, *Engines of Creation*, Doubleday, New York.
70. Lionel S. Penrose, 'Self-Reproducing Machines', *Scientific American*, 200(6) (1959), pp. 105–14. Videos of Penrose's models in action can be viewed at: https://www.youtube.com/watch?v=2_90hFWRoVs and https://www.youtube.com/watch?v=1sIph9VrmpM.
71. Homer Jacobson, 'On Models of Reproduction', *American Scientist*, 46(3) (1958), pp. 255–84.
72. Edward F. Moore, 1956, 'Artificial Living Plants', *Scientific American*, 195(4) (1956), pp. 118–26.
73. Dyson, *Disturbing the Universe*.
74. Freitas and Merkle, *Kinematic Self-Replicating Machines*.
75. Robert A. Freitas Jr, 1980, 'A Self-Reproducing Interstellar Probe', *Journal of the British Interplanetary Society*, 33 (1980), pp. 251–64.
76. R. A. Freitas and W. P. Gilbreath (eds.), 1982, *Advanced Automation for Space Missions*, NASA Conference Publications CP-2255 (N83-15348), https://en.wikisource.org/wiki/Advanced_Automation_for_Space_Missions.

77. Richard Laing, ‘Automaton Models of Reproduction by Self-inspection’, *Journal of Theoretical Biology*, 66(3) (1977), pp. 437–56.
78. Olivia Brogue and Andreas M. Hein, ‘Near-term Self-replicating Probes – A Concept Design’, *Acta Astronautica*, published online 2 April 2021, https://doi.org/10.1016/j.actaastro.2021.03.004.
79. R. T. Fraley et al., ‘Expression of Bacterial Genes in Plant Cells’, *Proceedings of the National Academy of Sciences, USA*, 80(15) (1983), pp. 4803–7.
80. Drexler, *Engines of Creation*.
81. Jim Giles, 2004, ‘Nanotech takes small step towards burying ‘grey goo”, *Nature*, 429, pp. 591.
82. Drexler, *Engines of Creation*.
83. ‘Nanotechnology: Drexler and Smalley Make the Case For and Against “Molecular Assemblers”’, *Chemical and Engineering News*, 81(48) (2003), pp. 37–42.
84. Salma Kassem et al., ‘Stereodivergent Synthesis with a Programmable Molecular Machine’, *Nature*, 549(7672) (2017), pp. 374–8.
85. A. H. J. Engwerda and S. P. Fletcher, ‘A Molecular Assembler That Produces Polymers’, *Nature Communications*, 11 (2020), https://doi.org/10.1038/s41467-020-17814-0.
86. Thomas C. Schelling, ‘Dynamic Models of Segregation’, *Journal of Mathematical Sociology*, 1 (1971), pp. 143–86.
87. ‘Parable of the Polygons’ is a game based on Schelling’s model: https://ncase.me/polygons/.
88. Schelling, ‘Dynamic Models of Segregation’.
89. Ulam, *Adventures of a Mathematician*.
90. Goldstine, *The Computer from Pascal to von Neumann*.
91. Robert Jastrow, 1981, *The Enchanted Loom: Mind in the Universe*, Simon and Schuster, New York.
92. https://www.gsmaintelligence.com/data/.
93. Jeremy Bernstein, ‘John von Neumann and Klaus Fuchs: An Unlikely Collaboration’, *Physics in Perspective*, 12 (2010), pp. 36–50.
94. Philip J. Hilts, 1982, *Scientific Temperaments: Three Lives in Contemporary Science*, Simon and Schuster, New York.
95. Dyson, *Turing’s Cathedral*.
96. John von Neumann, 2012 (first published 1958), *The Computer and the Brain*, Yale University Press, New Haven.
97. Von Neumann, *The Computer and the Brain*.
98. Robert Epstein, ‘The Empty Brain’, *Aeon*, 18 May 2016, https://aeon.co/essays/your-brain-does-not-process-information-and-it-is-not-a-computer.

99. Stan isław Ulam,'John von Neumann 1903–1957', *Bulletin of the American Mathematical Society*, 64(3) (1958), pp. 1–49.
100. *John von Neumann*, Documentary Mathematical Association of America, 1966. Many thanks to David Hoffman, the film's producer, for sending me the DVD in 2019. Now available to watch here: https://archive.org/details/JohnVonNeumannY2jiQXI6nrE.
101. Macrae, *John von Neumann.*
102. 'Benoît Mandelbrot – Post-doctoral Studies: Weiner and Von Neumann (36/144)', Web of Stories – Life Stories of Remarkable People, https://www.youtube.com/watch?v=U9kw6Reml6s.
103. https://rjlipton.wpcomstaging.com/the-gdel-letter/. Also see Richard J. Lipton, 2010, *The P=NP Question and Gödel's Lost Letter*, Springer, New York.
104. John von Neumann to Marina von Neumann, 19 April 1955, quoted in von Neumann Whitman, *The Martian's Daughter.*
105. Quoted in Dyson, *Turing's Cathedral.*
106. Quoted in ibid.
107. Email to author.
108. *John von Neumann*, Documentary Mathematical Association of America, 1966.
109. Ulam, *Adventures of a Mathematician.*
110. Quoted in Dyson, *Turing's Cathedral.*
111. Quoted in von Neumann Whitman, *The Martian's Daughter.*

EPILOGUE: THE MAN FROM WHICH FUTURE?

1. Frank, J. Tipler, 'Extraterrestrial Beings Do Not Exist', *Quarterly Journal of the Royal Astronomical Society*, 21(267) (1981).
2. Ulam, 'John von Neumann 1903–1957'.
3. Heims, *John von Neumann and Norbert Wiener.*
4. Von Neumann to Rudolf Ortvay, 29 March 1939, in von Neumann, *Selected Letters.*
5. 'Benoît Mandelbrot – A Touching Gesture by Von Neumann', Web of Stories – Life Stories of Remarkable People, https://www.youtube.com/watch?v=wu6vGDk5kzY.
6. Interview with author, 14 January, 2019.
7. John von Neumann, 'Can We Survive Technology?', *Fortune*, June 1955.

Image Credits

Pages 5, 7, 14, 40, 66, 67, 74, 98, and 283. Photos courtesy of Marina Von Neumann-Whitman.

Page 71. Stanislaw Ulam papers, American Philosophical Society.

Page 85. Courtesy of Los Alamos National Laboratory.

Page 106. The University of Pennsylvania archives.

Pages 131, 139. Alan Richards photographer. From the Shelby White and Leon Levy Archives Center, Institute for Advanced Study, Princeton.

Page 155: Courtesy of the University of Vienna archives.

Page 158. *Theory of Games and Economic Behavior*, John Von Neumann and Oskar Morgenstern, Princeton University Press.

Pages 194, 196. RAND Corporation.

Page 200: Courtesy MIT Museum.

Page 240. The Martin Gardner Literary Interests/Special Collection, Stanford University Library. Courtesy of Diana Conway.

Pages 249, 250, 251. Used with permission of Stephen Wolfram, LLC https://www.wolframscience.com/nks/<https://www.wolframscience.com/nks.

Page 256. Nils Aall Barricelli, 'Numerical Testing of Evolution Theories. Part I: Theoretical Introduction and Basic Tests', *Acta Biotheoretica*, 16 (1962), pp. 69–98.

Page 277. Briscoe Center for American History.

Acknowledgements

The list of those to whom I owe thanks is a long one, but there are two individuals without whom this book would not have been written at all. Marina von Neumann Whitman's encouragement provided some reassurance that I had not misunderstood her father altogether. Without Jeremy Gray's tireless help and support in the early stages of this project, I would have given the whole thing up as hopeless.

I am indebted to the following people who gave so generously of their time during the research, writing or editing of this book (in alphabetical order): Jim Baggott, Phil Ball, Daniel Bessner, Dennis Dieks, Jeremy Gray, Thomas Haigh, Tim Harford, Vaughan Jones (RIP), Shahn Majid, Michael Ostrovsky, Ulrich Pennig, Mark Priestley, Renato Renner, Katharina Rietzler, Andrew Wright and Costas Zoubos. In addition, Kenn Cukier, Steve Hsu and David Musgrave were early champions of *The Man from the Future*.

This book perches precariously on the shoulders of many giants. A comprehensive catalogue of their works would themselves fill a book, but readers plundering my bibliography will return with rich rewards.

Chris Wellbelove has supported my vision for this biography throughout some extraordinarily difficult times. The unerring instincts of Casiana Ionita, my editor at Penguin, have spared the reader from much unnecessary drudgery and toil. Any remaining drudgery and toil are entirely my responsibility. It takes a village to raise a book. This one was raised by Matt Hutchinson, Edward Kirke, Rebecca Lee, Imogen Scott, David Watson, Dahmicca Wright and Matt Young at Penguin Allen Lane, and Matt Weiland, with Steve Colca, Ingsu Liu, Don Rifkin, Rachel Salzman, Huneeya Siddiqui and Devon Zahn at W. W. Norton and Company. Thank you all for keeping the faith.

My wife and children have suffered with me and often because of me. I must also thank Mr Burnley, an inspirational English teacher, whose look of astonishment on hearing that I was going to study physics I have never forgotten. I hope this book is some recompense for that baffling decision. Finally, my mother, Sujaya Bhattacharya, always claimed I would write a book. Indian mothers are rarely shy about singing the praises of their offspring. I only wish she was here to embarrass me in person.

Index

Page numbers in **bold** refer to illustrations.